THE EXCLUSION EFFECT

The Exclusion Effect

How The Sciences Discourage Girls & Women & What To Do About It

KIRSTY DUNCAN

sh.
SUTHERLAND
HOUSE

TORONTO, 2024

Sutherland House
416 Moore Ave., Suite 304
Toronto, ON M4G 1C9

First edition, September 2024

If you are interested in inviting one of our authors to a live event or
media appearance, please contact sranasinghe@sutherlandhousebooks.com
and visit our website at sutherlandhousebooks.com for more information.

We acknowledge the support of the Government of Canada.

Manufactured in China
Cover designed by Jordan Lunn
Typeset by Karl Hunt

Library and Archives Canada Cataloguing in Publication
Title: The exclusion effect: how the sciences discourage girls & women &
what to do about it / Kirsty Duncan, MP, Ph.D.
Names: Duncan, Kirsty, author.
Description: Includes bibliographical references.
Identifiers: Canadiana (print) 20230593356 | Canadiana (ebook) 20230593399 |
ISBN 9781990823671 (hardcover) | ISBN 9781990823688 (EPUB)
Subjects: LCSH: Women in science. | LCSH: Sex discrimination in science. |
LCSH: Sex discrimination against women.
Classification: LCC Q130 .D86 2024 | DDC 500.82—dc23

ISBN 978-1-990823-67-1
eBook 978-1-990823-68-8

For Sven

Contents

CHAPTER 1

My Science Story

I SLOWLY RAN MY finger over the four numbers on the yellow page in the dark, dusty library stacks, where I had worked alone for six months, meticulously reading every book, journal, and newspaper article I could find on one of medicine's greatest mysteries. The year 1918 marked the onset of an unprecedented public health disaster: the great influenza (flu) pandemic, the deadliest in history. The Great Flu swept the world for two years, claiming the lives of over 50 million people—far surpassing the casualties of the First World War. Many victims were healthy, young people in the prime of their lives.

Despite the worldwide spread and alarm, the majority of those infected developed only mild flu symptoms: cough and stuffy nose, head and body aches, and a sudden high fever. The majority recovered and were largely back to normal in a week.[1] But some went on to develop pneumonia and other respiratory illnesses, and half of those victims died. People choked to death as their lungs became so swamped with blood, foam, and mucous that their skin turned blue and their lips purple.[2] The disease could also affect multiple organs and symptoms could resemble those of a chlorine gas attack, encephalitis, or nephrosis.[3] It was not until decades later that scientists were able to explain this phenomenon: a "cytokine storm" or severe immune-system reaction that could overload the body.

I was just starting my career as a scientist when I began researching flu in the bowels of the university medical library. The more I read, the more I wanted to unravel the secrets of the 1918 pandemic. I was insatiably curious. I loved puzzles and the thrill of discovery. Mostly, I wanted to help people and build a better world. Those drives are at the core of my love for science.

I read about the etiology, epidemiology, clinical manifestations, diagnosis, treatment, and prevention of influenza. I relished each new fact, statistic, and theory I learned about the 1918 flu. Along the way, I thought about the victims and their families, lost to history or buried for decades. There had been a shortage of medical personnel in those years as many doctors and nurses were serving in World War I. There were no effective treatments—no flu vaccines, antiviral drugs, antibiotics, or mechanical ventilation—only supportive and untested remedies. Medical science was at a loss to explain the pandemic or to provide sound advice to frightened people who witnessed those around them dying with astonishing speed.[4]

To slow the spread of the disease, governments implemented measures such as quarantine, placarded homes, closed public places, and regulated and enforced mask-wearing.[5] Individual citizens closed their doors to the outside world, communicated via letters, and some requested that telephones, then relatively new, be installed in their homes.[6] In desperation, families tried all kinds of preventive measures, from "vaccinations" to wearing mothballs around their necks to carrying hot coals soaked in sulfur throughout the house to ward off the threat.[7] Compresses, concoctions, and poultices made of bran, goose-grease, and turpentine were applied to the chests of the sick. Drinks of warm milk, ginger, pepper, and sugar were given to soothe the ill;[8] cough elixirs were administered to strengthen and heal the ailing.[9]

Chapels, gymnasiums, and jails were pressed into service as hospitals for the sick and the dying.[10] Still, the bodies piled up. The Philadelphia morgue could house about thirty-five corpses. But 500 people died in one day and 4,500 died in one week. Bodies wrapped in blood-stained sheets were stacked three and four deep in every corridor and room before, eventually, five more morgues were opened.[11] In Montreal, the demand for transporting the dead was so great that trolley cars had to be converted into hearses that could carry ten coffins at a time.[12]

In Canada, between 30,000 and 50,000 people died. Whole families disintegrated. Young adult victims left behind children who were forced into orphanages. Losses to businesses were staggering. Merchants lost their livelihoods because staff were absent with flu, and customers were too ill to shop. Pool halls, restaurants, and theatres all lost heavily.[13]

When I began investigating the 1918 influenza as a graduate student in 1992, almost seventy-five years after the fact, medical science was still unable to explain its cause, what made it so deadly, and why young people died in such great numbers. Influenza itself remained a major problem. Annual epidemics varied in severity, and it was impossible to predict when or where the next pandemic might occur, what viral strain might cause it, or how dangerous it might be. Moreover, history taught us that we were due for another fatal flu. The world's population had become much more vulnerable to the eruption and spread of infectious disease than in 1918 as more people traveled extensively and at greater speeds than ever before. Very few environments on Earth remained truly isolated or untouched.

I became determined to find the cause of the 1918 influenza. I wanted to be prepared for the next one. The more we know about pandemics, the greater the probability we can mitigate their deadly effects.[14] While I didn't know it then, I had embarked on a long, difficult road. I was a geographer, not a virologist—Strike one. I was young with virtually no track record—Strike two. And I am a woman—Strike three.

As a first step to finding the cause of the disease, I had to find lung tissue samples from its victims. Those samples might contain the virus. I contacted leading virologists from around the world, all of whom told me the same thing: no such samples existed. They were wrong. Some archival tissue samples would surface later.

I then tracked down information on two expeditions from the 1950s that had exhumed bodies of 1918 flu victims from the Alaskan permafrost in hopes that both the bodies and the virus would be preserved. Another dead end: no samples existed.

After six months of reading and painstaking archival research, I was stuck. I needed another tack. I decided to begin my own search for victims of the 1918 influenza, one that would take me to the frozen reaches of the planet. I began in Alaska and pored over thousands of death certificates. There were endless records but important details were often missing on the certificates: name, age, details of the illness, whether a doctor visited the patient, or the location of the interment. Moreover, I could not correlate individual burials with permafrost, as there was no such map for the state.[15] I was stonewalled once again, but the only way to ensure defeat would be to quit. Dead ends are common in science, and progress is built on failure. Yet we rarely talk about the missteps, the wrong paths, the struggles of science, and the courage, persistence, and resilience necessary to be successful. Making new discoveries often requires "a leap in the dark," with science proceeding in "fits and starts."[16]

Next, I wrote to leading Russian medical authorities about my idea and received no response, although years later I learned that American and Russian colleagues were considering a similar project. I continued my search and identified Svalbard, Norway, lying 1,000 km north of the mainland. I knew that the 1918 flu had sickened 370,000 and killed 7,300 people in the country, and I guessed that if people had traveled from Norway to the Arctic Archipelago, they might have brought the disease with them. It was just a hunch, but one informed by my training in geography.

I wrote to the Norse Polar Institute in Longyearbyen, Svalbard. It was interested but pointed out that there were no governmental records, as Svalbard did not become part of Norway until 1925. There were no hospital records, either, because the local hospital was destroyed during the Second World War, and there were no church records because the first minister did not arrive in Longyearbyen until 1920. But there were diaries, "Sigurd Vestbyes Dagbøker," kept by Store Norske Spitsbergen Kulkompani. I called the coal mining company, which informed me that it no longer had the diaries, but that the local schoolmaster and curator of the museum did. I tracked down the teacher who generously offered to translate them for me.

The diaries told the harrowing story of the first wave of flu in Longyearbyen, a community of roughly 200 people: "July 27th, 12 persons sick. July 28th, 14 more persons sick. 29th, 50 men to the doctor. 30th, Spanish flu has come! . . . July 31st, 25 men at work."

Then came a second wave. Almost seventy healthy, young people had boarded the ship *Forsete* in Tromsø in northern Norway, and all of them fell ill during the two-day crossing to Longyearbyen. Some of the men were so ill on arrival that they were taken directly to the hospital. Others were housed in extremely cramped barracks, allowing the flu to sweep through the small Arctic community once again. The diaries report: "This season's last ship. 7 men died and were buried October 27th 1918 in the churchyard at Longyearbyen." The miners were between nineteen and twenty-eight years of age.

I should have been elated, but my excitement was tempered by the fact that the 1918 influenza was no longer just a scientific puzzle. Now it had seven faces, and they were just as young as I was. I wondered about the young men—who they were, whether they had dreams of a new beginning and a new life, whether they were afraid, and who they left behind.[17]

Finding the diaries was only the beginning. I needed to know whether or not the graves were marked or had been disturbed. What were the Norwegian burial practices in 1918? Archaeologists, historians, Norwegian funeral directors, the Lutheran Church, and governmental authorities told me that the bodies would have been placed in simple wood coffins without embalming fluid and buried two meters underground. While this information was encouraging, Svalbard was "no-man's-land" in 1918, and it was impossible to know if these practices had been followed.

The next question was whether the ground temperatures were cold enough to allow biological preservation. I reached out to a permafrost specialist at the Geological Survey of Canada who worked out that at a depth of 1.0–1.5 meters, ground temperatures would have ranged from -10.0°C to -4.0°C over almost eight decades and were considerably colder than a standard morgue.

After two-and-a-half years of work, I had put together an ideal case. I had found the final resting place of seven miners. Their graves were marked by six white crosses and an obelisk. The graves were undisturbed and the burial practices, depths, and ground temperatures suggested that the bodies and fragments of virus might be preserved. My plan was to take small tissue samples from the miners to characterize the 1918 flu virus, but there were difficult ethical considerations. While the victims might hold the secrets to the 1918 flu, I struggled with the idea of disturbing their final resting places. At last, my father said to me that if his body held the answers to a deadly disease, he would hope that someone would come along and unravel those secrets. I decided to pull together a multidisciplinary, multinational research team with whom I could collaborate.[18]

A few years out from completing my PhD, I believed I could invite a group of leading scientists from around the world and that we would all work together to solve the mysteries of the 1918 flu. The team members would come from geography, geology, medicine, medical archaeology, microbiology, pathology, and virology, and, ultimately, representing four countries: Canada, Norway, the United States, and the United Kingdom.

In theory, this impressive group of people and disciplines should have led to an inspiring collaboration to find the virus, with each team member bringing valuable expertise. However, the virologists saw themselves as alpha scientists and complained that "a geologist could never understand the work of a virologist." They failed to understand that without the geophysicist, geologist, and pathologist, there would have been no work for the virologists. There was also bitter competition among the virologists themselves, each seeking the "holy grail of influenza research." On top of this, countries that had often voted *en bloc* on lab agreements, memorandums of understanding, and publishing, could also be competitive with one another.

Regardless, I maintained a singular focus. My next step was to request permission from the governor of Svalbard to undertake the exhumations. In return, I promised that Norwegian scientists would be involved in the planning and undertaking of the work.

Shortly thereafter, a stranger informed me that he had been appointed as a member of my research team by the Centers for Disease Control (CDC) in the United States. Why was CDC openly discussing my project? Why had it invited a new team member? And why was this stranger making recommendations on a project he knew nothing about? I let him know that I did not appreciate decisions being taken without me. He replied, "Rest safely, no one will take it away from you. There is no chance of your losing control." It struck me as an unusual thing to say, unless it was an attempt to hide a desire to assert control. At any rate, it became a familiar refrain. Various team members would try to gain control of the leadership, funds, and ultimately, the release of the results. As an early career researcher and an athlete all my life, I could push through anything. I had a goal and the resolve to achieve it. I would stick to my ethics and do my utmost to keep a team of shifting international alliances on task and moving forward.

Four years into the work, the surviving family members of the miners said "yes" to my very sensitive request to undertake the exhumations, and the governor invited me to Svalbard. It was my opportunity to meet the community—the governor, the church, the Norse Polar Institute, the teacher, schoolchildren, and the people of Longyearbyen—and build relationships and trust. I assured them that if I were allowed to do this work, it would be done safely and with the greatest dignity and respect.

While I worked, a new virologist claimed credit for my project, and my university thought it best for me to go to the media and stake a claim. The ensuing media blitz motivated even more virologists to seek involvement, with one man demanding, "Young lady, you know who I am, don't you?" I would hear more condescending, paternalistic words about ownership and control from him. "My young lady, you're a neophyte in this area. None of us want to take what you achieved."[19]

Scientific planning continued with meetings at the University of Windsor in Canada, the CDC in the United States, the National Institute for Microbiology (NIMR) in the United Kingdom, and the University of Toronto. At a Windsor meeting, the team had agreed to look for tissue samples in Svalbard, although we would still need permission from Norway's Directorate of Cultural Heritage. I continued to share all the information with the team. Some team members had been tasked with searching for archival samples as an alternative to exhuming bodies; I noticed that these researchers either drifted off or refused to share information.

A month after we received the required permissions from the Directorate of Cultural Heritage, Dr. Jeffery Taubenberger, an American virologist, and his team independently published their significant work on a successful analysis of fragments of RNA from archival tissue samples. It was a remarkable achievement and the very first glimpse into the 1918 virus. We invited Dr. Taubenberger to our team's next meeting at the CDC to learn about his work, as we faced a decision on whether or not to proceed to Svalbard.

At the meeting, the CDC made it clear that it was no longer interested in archival material. That surprised some scientists on the team, but by then I had grown accustomed to the organization's routine reversals of position. There were strong egos throughout our growing team. The stakes were high, and the politics were turbulent—often descending into "double-dealing," feuding, and "power games" among the world's best.

After much discussion, the team agreed that our project would only involve autopsy material as opposed to archival samples. We would also do an initial ground penetrating radar (GPR) study of the cemetery and proceed to the exhumations only if the GPR data warranted it.

After the meeting, Dr. Taubenberger and I met to discuss our research. Like me, he was an early career researcher. He shared his difficulties in publishing his results, how he felt marginalized by the "influenza Mafia" of more established scientists, and how he trusted virtually no one who sat around the table at our meeting. At the request of our team, I invited Dr. Taubenberger to participate in the Svalbard project. He said, "Okay, boss. Friends." We shook on it.[20]

When commiserating with Dr. Taubenberger, I did not talk to him about the difficulties of being a young woman leading a complex, international research project with male team members who were double my age. I didn't tell him about the official who, while driving me back to my accommodation, drove past it, locked the car doors, and put his arm across my chest. I didn't express the absolute fear I felt at that moment or tell him how I always tried to put a door between me and another team member who could not stop looking at my legs during conversations.

Like any scientist, I just wanted to do my research and I was tired of questions that were meant to undermine me. Team members asked, "How do you want to be treated, as a woman or as a scientist?"

I flatly responded, "The two are not mutually exclusive. As your colleague."

One scientist shot back, "But you wear short skirts."[21]

I did not comment on their attire, yet they felt free to question my hair, my dress, my passion for teaching dance and running marathons—which they considered unserious pursuits—or anything else that, in their minds, served to highlight how different I was from the stereotypical white, middle-aged, male scientist in a white lab coat.

We needed funding. The project had never before been at a stage where I could apply for grants, and I could not keep contributing to it from my own pocket. I had already spent over $20,000 on faxes, telephone calls, conference proceedings, and travel to get us this far. Two of the virologists offered to draft grant proposals: one application to the National Institutes of Health (NIH) in the United States and another to Roche, the pharmaceutical company. I gratefully accepted. I was busy liaising with government officials and overseeing the planning of the GPR study.

I ended up having to fight to be a coprincipal investigator on both applications after one of the virologists asked me for a letter of collaboration on "his" research project. I remembered his earlier words: "None of us want to take what you achieved." It occurred to me that the funds would eventually go to their institutions and their countries and could be used as a bargaining chip in future discussions and negotiations. It had become clear that we needed a memorandum of understanding to prevent future academic squabbles and to make sure all involved would receive due credit.[22]

The CDC then announced that it was pulling out of the project altogether. Only two months prior, it had committed to funding the GPR study and providing a Biosafety Level 4 (BSL 4) lab, one of only a handful in the world. This complete reversal put our funding from the NIH in jeopardy. I needed to find a new lab quickly.

Despite the massive challenges, the project came together. I had permission from Norway, we had funding from the NIH and Roche, new team members coming on board, a new BSL4 lab, and plans for the GPR study that were progressing well. We were careful about all our preparations yet one virologist wanted to skip the GPR and quickly exhume bodies "before they all wake up." He even recommended that we try to do the work anonymously in Svalbard—on a barren hillside in a community of 1,200 people. I ignored him and continued to prepare.

I booked accommodations, cars, GPR equipment, and transportation, and arranged for equipment insurance and meetings with the governor's office, the Directorate of Cultural Heritage, and the church. I reconfirmed permissions to undertake the exhumations. I focused on reducing the risk of infection, the need for rigorous safety precautions, respect and dignity for the deceased and their families, and limiting damage to the cemetery, a protected cultural site.[23]

At last, we arrived in Longyearbyen. I made my round of visits to the community. I cared about the people and I hoped to earn and maintain their trust. I never wanted to be that absentee, foreign researcher. Afterward, I joined our team members leading the site survey in preparation for the GPR study. The radar used radio waves to detect buried objects in non-metallic material; it had previously been used to locate land mines, murder victims, and even a lost flying squadron buried deep in the Greenland ice sheet. I watched as specialists dragged the GPR equipment one hundred times across the bumpy, frozen tundra and waited with bated breath for images to slowly appear on the computer screen.

We collected a huge amount of data that almost maxed out a hard drive and required months to interpret. We then laid a wreath, and I said goodbye to the community—perhaps for the last time, perhaps until exhumations.[24]

Not long after, I was invited to the NIH in the United States as one of two coprincipal investigators on our grant application. We were to discuss risks and benefits, our progress, and our plans for the expedition. A "select group of individuals," including former team members and current supportive ones such as Dr. Taubenberger, would rule on the project.

A day after receiving the invitation from NIH, I received a fax from Dr. Taubenberger who said he was withdrawing from the project. Astounded, I called him. He explained that his boss had forced him to send the fax and he apologized. At the NIH meeting, he changed course again and suggested to everyone present that the project should not go forward. He said it was a very long shot.

I was shocked, and I still had to present our GPR data. I explained that the radar did not show coffins, but rather disturbed ground that extended to a depth of two meters. Based on temperature records over the previous eighty years, if there was any material beneath the surface, it was likely frozen.

Three days before I traveled to our team's next meeting at the NIMR in London, England, Dr. Taubenberger called. "I hope you won't think any malice was intended," he said.

He went on to explain that a retired, out-of-state pathologist named Dr. Hultin, who had been involved in a 1950s expedition, had offered to return to Alaska to exhume flu victims and send samples to Dr. Taubenberger. Dr. Hultin had made his preparations in one week, asking permission only of elders. He had assumed there was no live virus and excavated in the open air with the help of four young men, using his wife's pruning shears as autopsy tools. He sent four packages of samples to Dr. Taubenberger via courier and post. He also provided a letter explaining that Dr. Taubenberger was not involved in the exhumations.[25]

In stark contrast, I had gone through a first permission process of sixteen months and a second one of seven months. My team had spent four and a half years planning our project, with one and a half years spent on the biosafety protocols alone. We justified our safety protocols to the experts at NIH, some of whom would congratulate the casual foray to Alaska despite its serious lack of safety precautions. We always felt that we had to do things right. I would not have done them any other way. The American team, by contrast, just wanted to get results.

On our end, plans for the Svalbard exhumations had finally shifted into high gear. We provided the governor with a risk analysis; a fieldwork plan; an infection, ethics, and cultural heritage plan; and a media plan. We promised that afterward the cemetery would be fully restored, with no visible trace of our expedition. My focus had always been the safety of our team and the nearby community, as well as respect and dignity for the deceased and their families. The hospital in Longyearbyen and the nearest hospital on the mainland knew of our work and were prepared to respond if there was accidental exposure. Only a minimum number of people were allowed in the tent over the gravesites, and everyone had to sign a logbook every time they entered and exited the site. All involved in the exhumations would take an anti-influenza drug as a precaution, as well as a new experimental drug not yet on the market. The final biosafety protocols were exhaustive.[26]

Having met all of Norway's requirements, we were on our way back to Svalbard. It was the culmination of six years of work, with possible answers to an eighty-year mystery.

My first job was to visit the community, renew relationships, answer questions, and allay any concerns anyone might have with the exhumations. The next job was the construction and preparation of the site. Two containers with seventeen tonnes of supplies that had been packed in England arrived in Longyearbyen and were transported to the cemetery. Protective walkways were put in place to protect the fragile Arctic tundra, as one misplaced footstep could have left a mark for a decade or longer. The white crosses over the graves were carefully removed from the ground and carried downhill for safe storage. A mobile surgical unit and a -70°C freezer were winched up the steep slope.

Next, the surface layer was carefully cut, removed, and stored so it could be returned to its exact location. With that, we began digging in the permafrost. A half-meter down, one of the scientists struck a wood surface in the active layer of the permafrost (which freezes and thaws seasonally). We immediately stopped work. A small team went up to the tent to investigate, while the rest of us huddled in the work cabin and waited for information and the world's media awaited news.[27]

I had some experience of the media. I'd been questioned on whether or not I was a scientist, whether or not I was serious, or whether or not I was aware of the knives aimed at my back. I was more than aware. I had lived it for four years. I had pushed through sexism, sharp elbows, and treachery. While the men were given their titles in news reports, I remained "Ms. Duncan, 32 years of age, and a divorcee." I was even asked if I would lie on a sports car in front of a cemetery for an article and photograph. Despite it all, I had stuck to my ethics, kept my promises to Norway, and stayed above the ever-present dark side of science.

After a few short minutes, a radio call confirmed that the exhumation team had found a coffin. Hushed silence. In a short time, two more coffins were found. Given what looked like good radar images, it was surprising to find the bodies in the active layer. Were they buried more deeply back then, had they risen through time, and might they have come closer to the surface later?

Eventually, six coffins were opened, as per my permission from Norway. The bodies were wrapped in newspaper dated 1917, confirming that these were, in fact, the miners from the diaries.

I had waited six long years for this deeply emotional moment. I hoped that these men from nearly a century ago would reveal their story and their fight with the virus. I thought of their families.

We had come to Longyearbyen to obtain soft tissue samples, and we had done it safely and ethically. In total we had collected over one hundred samples from the bodies of six miners. The precious cargo was packed to prevent mechanical and temperature damage and flown out on two separate flights to the NIMR in London, England. The tents and everything else were removed and the soil was replaced. The crosses were put back in place along with the curb stones around the graves. The shipping containers were removed and funds were set aside should the cemetery ever require further cleanup.[28]

The NIMR had promised to test for live virus on receiving the samples. Once the samples were deemed safe, they were to be shipped to Canada, Norway, and the United States. But the NIMR broke its agreement. It tested for infection and PCR screening in parallel. That meant our British colleagues were giving themselves a jump on the other countries. They appeared to be competing with the rest of the team. Moreover, NIMR refused to share information with us despite repeated requests.

It wasn't until eight months later that I learned scientists at NIMR were receiving "signals" from the brain, kidneys, liver, and lungs. At least it was good news. Seven years of work and there were results.

While I pushed for more information and for the samples to be returned to Norway, one of the British virologists on my team, backed by other British members and an American member, invited others from our team to an influenza conference in London, England. I was left out, along with our Norwegian lead. When confronted, the British virologist wrote that "this is a scientific meeting . . . and not about Spitsbergen (Svalbard)."

Two months later, on a Thursday night, I received calls from the British media asking if I was part of Professor Oxford's Spitsbergen expedition. I made it clear that I had founded the project and was the project leader and that he was one of my team members. The media informed me that the results of my expedition would be released on Monday in Britain.

Professor Oxford was stealing from Norway and he was stealing from the team. It was academic piracy.

The next day, I called the *Toronto Star* who interviewed me and the team. In the meantime, I received a copy of the British press release. It stated, "The symposium will provide the first opportunity to hear the data from the Spitsbergen expedition undertaken by Professor Oxford and his team." I booked my flight to Britain.[29]

On Monday morning, I muscled my way into Professor Oxford's conference. I told him I was there to get credit for Norway and our team. If he did not recognize us both, I would stand up and set the record straight. Amazingly, he had invited Dr. Taubenberger and Dr. Hultin, whom he was now courting. Our American team member said to me, "You've won me some money." He had actually bet others that I would show up to the conference. It was surreal, but he knew I would not go down without a fight.

Four members of my research team spoke at the conference, and one member presented the team's results. He noted, "There is a young lady in the audience who kinda got the ball rolling in Longyearbyen." He then put up a slide with all my team members' names and briefly described each person's contributions. And there was a note thanking Norway. Excellent!

But I was not done. My presence at the UK conference forced the cabal to issue a new press release, this time without mention of Professor Oxford and his team. Earlier, he had brought his own journalist, his daughter, to Longyearbyen against the team's guidelines, and secretly recorded a radio commentary.

During the conference, I explained to its funder that this had been a seven-year undertaking and a team effort, that the contributions of all members were necessary, and that Norway had given us a great gift. Later, at a press conference, the funder ordered Professor Oxford to do the right thing. I sat, glaring, mouth pursed, and arms crossed in the front row of the press conference. He had no way out. I had come to England to get credit, the credit that was owed to Norway and my team, and I did exactly that. Professor Oxford credited Norway, the team, and me.[30]

In the years that followed, I made sure the cemetery was healing and that the governor was pleased with the recovery. It became more and more difficult to get information out of NIMR. In 2000, NIMR suggested that Dr. Taubenberger, who had worked behind our team's back with Dr. Hultin, be involved in a blind analysis of our samples. The Canadian lab reported that it was unable to confirm the results, but that the NIMR data looked good.[31]

It wasn't the best outcome we could have hoped for but, again, negative findings are progress in science, and we had made other important contributions. Despite innumerable challenges, we developed new safety protocols to protect our research team and the nearby community, and we developed new techniques to safely exhume bodies and remove tissue that might contain the virus.

I'd had a bold idea that caught the imagination of the world's best and might have given us answers to a deadly killer. It was unfortunate that egos, politics, and the rush for personal acclaim hurt the project, the team, and threatened the science. As difficult as my journey was, I pushed through because of my belief in the project. I wanted to be ready for the next pandemic, whenever or wherever it began. Also, because of my commitment to Svalbard and its people, and my doubts that certain members of the team would honor the permissions I had been given.[32]

The expedition raised important questions about epidemiology, ethics, public health, and the rights of subjects. It also raised questions about age, gender, and privilege in science, which at the time were largely taboo subjects. I had to fight hard to include any reference to gender in my academic book on the expedition, as it was considered a fringe element not relevant to the serious study of science.[33]

* * *

Would my journey have been different if I were a man and older? Probably. But one thing is certain: because of the way I was treated, I went on to devote my life to making science better by inviting everyone into the classroom, field, and lab; and by inspiring, mentoring, and supporting women to achieve their science dreams.

Science holds real promise for people and our planet. It is our best hope to find solutions to improve health, address biodiversity loss, combat global pandemics, respond to natural disasters, and tackle the climate emergency. It has the potential to develop new technologies, jump-start economic growth, and change all our lives for the better. But science also matters for its own sake. At the heart of science is research, which is about adding to the sum of the world's knowledge—it is about learning and discovery. Most importantly, science belongs to everyone.

Fewer women than men pursue STEM (science, technology, engineering, and mathematics). As we'll see, the reasons are multifaceted. They include negative experiences in the classroom and on campus, abuse, discrimination, harassment, implicit bias, negative stereotypes, a lack of role models and social support, and a "chilly climate" in the profession based on the mistaken belief that women "do not belong in" or "cannot do science." All of these take their toll over time and discourage young women from entering STEM careers—even with its promise of better pay. At present, it seems that many women want more from their careers than those in STEM currently offer. That has to change.[34]

Every girl who dreams of pursuing a life in science should know that there is an infinite number of ways to do so—that there is a place for her and that chasing that place is her birthright. Every girl, when asked the question of what she wants to be when she grows up, should know that answers like space exploration, robotics, and science are well within her reach. Ensuring those futures is the purpose of this book.

CHAPTER 2

The Brain, Ability, and Born to Explore

BEFORE SCIENTISTS COULD LOOK inside the human brain, there was a belief—from the eighteenth century onward—that biology determined destiny. This dangerous, simplistic notion became conventional wisdom and led to different educational standards for women and for men.[35] Male-centric theory postulated that because they had bigger brains, men were fated to be brilliant, men would become scientists, and, ultimately, Nobel Prize winners. Women were destined to be the emotional, nurturing companions of these men, supporting their careers and giving care to their children.[36]

It was only within the last fifty years that researchers developed sufficiently powerful scientific tools to make the invisible visible in a living human brain. But instead of using these tools to question and challenge ancient dogma, researchers used them merely to confirm what was already treated in society as fact: that there are physical brain differences in the brains of women and men.

Cognitive neuroscientist Gina Rippon, Emeritus professor at the Aston Brain Centre, Aston University, shatters this self-serving dogma in *Gender and Our Brains: How New Neuroscience Explodes the Myths of the Male and Female Minds* (2019). Dr. Rippon argues that brains are not fixed as female or male at birth, but are highly plastic. They change constantly throughout our lives and are influenced by the gendered world in which we live.[37] Once differences in brain size are accounted for, the "well-known" sex differences in key structures effectively disappear. Science is showing us that the human brain is no more gendered than the heart, kidneys, or liver.[38] The transformative takeaways from Dr. Rippon's research are that girls' and boys' brains are plastic, and that biological sex alone does not determine ability and achievement. All children have potential. Dr. Rippon's book represents a seismic shift in thinking, one that parallel social science trends support.

The new techniques in brain imaging made it possible to look at the neural networks of developing brains *in utero*.[39] From the moment of conception, a baby's brain grows very quickly: at about 250,000 nerve cells per minute.[40] At birth, 100 billion nerve cells are already in place, even though a newborn's brain weighs only about 350 grams or about one-third of what an adult brain weighs, 1,300–1,400 grams.[41]

Baby boys tend to have larger brains than girls, but this difference disappears when body size is considered, and there are very few, if any, structural differences in the brain at birth once birth weight and head size are factored in.[42] At this point in a baby's development, there is nothing that will reliably distinguish a girl's brain from a boy's brain.

One challenge of studying any possible differences between infant girls and boys is that the newborn brain grows very rapidly, at about one percent per day for approximately the first ninety days, at which point the brain has doubled in size. A second challenge is that even if structural differences were to be found between the brains of infant girls and boys, any functional implications would be unclear.[43]

Rippon tells us that baby brains are not just unfolding according to some preordained genetic code. They are highly mouldable or plastic: they can change activity in response to internal or external factors by reorganizing connections, functions, and structures.[44] Almost anything can shape us, for good or bad, in the early baby and toddler years—first words, television programs, toys, and more. The brain will continue to change through experiences such as learning to walk and developing independent thought, throughout our lives.

By about five months, babies grasp simple math and physics. There is no gender gap in performance. In fact, as early as two days after birth, babies can tell the difference between small and big numbers by matching short bursts of beeps to pictures with a few smiley faces versus long bursts of beeps to pictures with many smiley faces. And by two to three months of age, they express surprise when a ball that has rolled into a tube does not roll out the other side.[45] Babies not only understand what an object is but also recognize that it cannot occupy the same position as another object and that it can't go through a solid boundary.[46]

Harvard Professor Elizabeth Spelke found "no evidence for a male advantage in perceiving, learning or reasoning about objects, their motions, and their mechanical interactions" after she reviewed thousands of studies conducted over three decades.[47] To be clear, Spelke found no difference in aptitude for math and science.

Spatial cognition or the ability to understand the relationship between two-dimensional or three-dimensional objects has long been treated as a biologically based skill that is needed for science. Males seem to have more of it.[48] However, when this supposed difference is measured in different ways, females perform equally well or better. We also need to remember that brains are plastic and can be trained over time. What you start with isn't what you end up with.[49]

By age four, the brain has quadrupled in size with 700 new neural connections forming every second. By age six, the size of the child's brain is about 90 percent of an adult's brain.[50] While each brain is unique, adult women and men have the same brain structures in much the same proportions. Women and men also carry out emotional, language, memory, and spatial tasks using very similar processing.[51]

It's past time to move beyond the myth and stereotype that categorize women as right-brained—emotional, nurturing, and good at verbal tasks—and men as left-brained—logical, rational, and good at spatial tasks.[52] These beliefs have done enough damage to women scientists over generations.[53]

Rosalind Franklin University neuroscientist Lise Eliot and her collaborators undertook a comprehensive study of three decades of research. They looked at hundreds of the largest and most frequently cited brain imaging studies of thirteen measures of alleged differentiations on the basis of sex and found "no universal, species-wide brain features that differ between the sexes."[54] Women are not from Venus and men are not from Mars, as was once famously postulated by relationship counselor John Gray.[55] Rather, the brain is unisex, like other organs, including the heart, kidneys, liver, and lungs.[56] Importantly, the variability among women's or men's brains is greater than the variability between them: there could be just as much difference between two women's brains, as between a woman's and a man's brain.[57]

But, again, brains are not isolated biological organs cut off from the world. The world influences the brain in interesting and important ways. Brains are also predictive and sensitive to social context. The brain just doesn't receive and process information, as was once believed; it picks up rules and looks for shortcuts and predictability.[58] And because humans are a social species and dependent on cooperation to survive and thrive, it should come as no surprise that babies are born ready to respond to their social environment. Within hours of birth, they react differently to faces, and within days, they respond differently to their native language.[59]

Our brains also guide us away from things that make us feel bad, such as lack of belonging, rejection, and social mistakes, and drive us toward things that make us feel good. In fact, the parts of the brain that are activated when someone makes a mistake are the same parts that are activated by real pain.

Clearly, there is a close interplay between nature and nurture. We are not stuck with the brain we're born with. Our personalities, skills, and futures are not predetermined. Our brains are highly individualized, adaptable, and teeming with potential, which is why environment and experience matter.[60]

* * *

From ancient times, Western cultures trained children to do different jobs in families and societies. There was a strong focus on gender roles and on predicting the sex of a baby: dangling rings, food cravings, heart rate, and the position of a "baby bump" were all used to foretell whether a growing fetus would be a girl or a boy. Yet on nearly every continent and throughout history, thriving cultures recognized more than two genders. Examples include the Bakla in the Philippines, Hijra in India, Mahu in Hawaii, and Two-Spirit people across Turtle Island.[61] Today, we remain focused on the sex of a baby: with the advent of fetal ultrasound, we see the rise in popularity of "gender reveal parties," ranging from ones with classic cakes with pink or blue fillings to another where partygoers shot at "girl" or "boy" targets packed with highly volatile Tannerite. At the latter, a blue cloud exploded and sparked a fire that ultimately burned 47,000 acres in Arizona.[62]

Evidence shows that pregnant mothers describe their baby's movements *in utero* as "strong" and "vigorous" when they know that they are having a boy. There is no such use of gender-contrasting terms when mothers do not know the sex.[63]

Sex refers to the biological and physiological characteristics of females, males, and intersex persons and is typically assigned at birth, usually based on external anatomy. *Gender* refers to the characteristics that are socially constructed, because of norms, roles, and behaviors.[64] Gender is also hierarchical, with males viewed as the dominant and powerful sex, and gendered relations ranked and evaluated according to a standard of masculine norms and behavior. *Gender identity* refers to a person's concept about her, his, or their gender and is a deeply held, internal sense of self that is typically self-identified.[65]

Although it is increasingly understood that gender is not binary but rather a spectrum and that people have their own words to describe themselves, babies largely continue to be welcomed into a highly gendered world: stereotypic pink or blue blankets, clothes, rooms, teddy bears, and toys.[66] The pink and blue worlds of girls and boys are a rather recent phenomenon. For centuries, children wore white dresses until about six years of age as a matter of practicality: dirty white cotton was easily bleached. Pink, blue, and other pastels did not appear until the mid-nineteenth century; pink and blue did not come to specify gender until just before World War I.[67] Blue was originally the color for girls, as it was seen as dainty and refined and as Professor Rippon points out, has also been associated with the robes of the Virgin Mary. Pink, initially, was a strong color and therefore appropriate for boys. The opposite, pink for girls and blue for boys, became firmly established in the 1940s by American manufacturers and retailers, with baby boomers growing up dressed like their mothers and fathers.[68]

In the 1960s, the women's liberation movement triggered a backlash and ushered in anti-fashion, anti-feminine, and gender-neutral clothing trends, which remained popular until roughly 1985, when prenatal testing allowed expectant parents to shop specifically for girls' or boys' clothes and toys.[69] When I was growing up, there was little pink. But I liked the color and chose to wear it and as a result was threatened in the school halls and washrooms. The color was seen as a symbol of being "gay" and "weak." I dug in my heels and chose to wear the color more to stand up for those who were being bullied.

Today, pink is a brand associated closely with girls. Parents who aim to be gender neutral will often buy their daughters clothes typically marketed for boys. Rarely do we see the reverse. The belief that parents need to buy their sons boyish clothes perseveres.

Parents and caregivers also use different words to describe girls and boys and their actions, even for the exact same behavior. And they provide gendered toys that can reinforce hobbies and traits that are already socially preassigned to females and males.[70]

Boys are more likely to be given construction blocks such as Lego, encouraged to play sport, and engaged in activities that have strong spatial elements. The so-called biological difference in spatial cognition that supposedly gives males the advantage in science is perhaps better explained by differences in spatial experience.[71]

While all this is happening, we cannot forget that babies are busy soaking up the world around them and are extremely sensitive to social cues.[72] Children start to conform to "stereotypes" of females and males very early in their development. At two to three months of age, babies can discriminate between female and male voices.[73] Between eighteen and twenty-four months, toddlers develop the ability to recognize and label stereotypical gender groups, such as girl, woman, and feminine; and boy, man, and masculine.[74]

Around two years of age, girls and boys start to diverge in their toy preferences; by three years, they start to figure out "I am a girl" or "I am a boy," and they become very invested in their "category."[75] This, in turn, prompts a child's preference for pink or blue, different activities, clothes, and toys. And by ages five to six years, they are often rigid in their gender preferences and stereotypes.[76]

Attitudes, behaviors, and expectations toward girls and boys remain stubbornly different. It's how we talk to girls and boys, how we make eye contact and gestures, and whether we expose them to arts, athletics, or play.[77] A quick Google search of girls' activities recommends drinking tea, getting glammed up, or playing make-believe or princess. For boys, recommendations include getting muddy, going on a scavenger hunt, or shooting stuff. I wish I was kidding!

While adults and peers continue to model and expect certain behaviors and focus on differences, they help to consciously or inadvertently perpetuate a gendered world.[78] Pervasive advertising further reinforces social conventions. A global survey by the Geena Davis Institute in partnership with LEGO in 2021 showed that three-quarters of parents would encourage their sons to play with Lego blocks, compared to only a quarter who would encourage their daughters. Lego has since committed to removing gender stereotypes from marketing and products.

Most differences in behavior between girls and boys stem from the environment, and they can be counteracted through equal opportunities, multiple forms of play, and toys.[79] Parents, teachers, and caregivers should focus on activating and exercising as many circuits as possible in the developing brain, much like cross-fit training tackles conditioning, functional movement, and strength training, to give children the best start possible for whatever they choose to do later in life. Parents and carers can also help children as they begin to identify gender divisions. Adults can revise stereotypes and let children know that girls can play hockey, play soccer, and be scientists.

The human brain should be seen as just that—human. It is now abundantly clear that the brain has infinite possibilities, that what babies and children are exposed to early on in their lives matters profoundly, and that it's time to stop pigeonholing children into binary, stereotypic pink, and blue environments that curb exposure, limit experience, constrain opportunity, and hinder futures. Based on the best available science, girls absolutely have the brains and the aptitude to be successful scientists.[80]

* * *

Babies and toddlers are interested in everything. They crawl and climb to discover, test, and understand their environment. They love to play with water, whether in puddles, sinks, or tubs, and they let the rain fall through their fingers. My mom loved to share how once as a baby, I pulled out all the pots and pans from the cupboard and clanged away on my new instruments; and how another time, I got into the fridge, pulled out a loaf of bread, happily tore it up, and threw the remains around the kitchen. Children learn by hitting objects together to test different sounds and by banging, dropping, and stacking things up and watching them fall. Babies and children are like little scientists.[81] They are constantly discovering, testing, hypothesizing, problem-solving, and even drawing causal conclusions.[82]

Researchers at John Hopkins University created an optical illusion where a ball rolled downhill and appeared to go through a wall. Seventy percent of the eleven-month-olds further investigated by banging the ball, putting it in their mouths, or rotating it in their hands. In stark contrast, only twenty percent of infants tested the ball's solidity when the wall stopped the ball. What is significant is that when the object defied their expectation, infants paid attention, explored more, and tested the unexpected.[83]

Studies over decades show that babies look longer when they come across something surprising, such as a toy car that magically appears to float across a gap in a track or pass through a wall, and that their interest is demonstrated through physical changes like an increased heartbeat.[84]

Babies learn so fast that they transform their worldviews every few months and embrace new ways of understanding their existence.[85] Berkeley child development psychologist Alison Gopnik, in her 2011 TED talk, described how babies even make complicated calculations with conditional probabilities that they revise to figure out how the world works. And by age four, children who are just learning how to count will sometimes test five hypotheses in just two minutes; and they are better at finding an unlikely hypothesis than adults.[86]

Children not only are like little scientists but also exhibit a key characteristic of researchers: namely, curiosity. A survey of around 400 members of elite U.S. scientific societies, such as the National Academy of Sciences, lists honesty and curiosity as the most important traits underlying excellent science.[87] Children— girls, boys, non-binary, and gender diverse—are born curious, learning through every experience and social interaction they have. (I write from a feminist intersectional perspective, and I want to be as inclusive as possible. I am not sure that we always have the right language, but I do know that it's important that we all keep listening and that we avoid generalizations, as people have multiple identities and experiences. Moreover, language continues to change. Some terms used today are different than those in the past, and some people use less commonly used terms to describe themselves. What matters is that we properly use gender identity terms and pronouns and that we recognize and respect people as individuals.)[88]

Girls actively explore the world around them. They crawl into small spaces, they climb on the furniture, and they like to go to the beach, the forest, and the marsh. They like the colors of flowers, and they like to feel the texture of grass, moss, and tree bark. Girls, like boys, non-binary, and gender-diverse children, are born to explore, and their interest in the world around them is limitless. Giving children the opportunity to be curious and explore their environment is important for their development and well-being.[89]

As children develop language, they add a powerful tool to their learning toolbox: asking questions.[90] A survey of 1,000 British mothers with children aged between two and ten years showed that they are asked more questions every hour than doctors, nurses, or primary school teachers and that four-year-old girls asked the most questions.[91] More recently, British researchers polled 1,500 mothers and fathers across the United Kingdom in 2017 about the frequency and nature of their children's questions. They found that curious children asked over seventy questions a day, and parents struggled to answer half their children's questions.[92]

The point is that children inundate parents and carers with questions about why and how. Why is the sky blue? How do birds fly? Where did I come from? How was I made? Girls, boys, non-binary, and gender-diverse children all ask questions. Questions are the essence of science. Girls and boys are not only are curious but also love to discover, explore, and ask questions, and these are the building blocks of science. They are natural scientists and by five years of age, all children are great problem solvers and are generally ready to learn as they enter kindergarten.

I have always made it part of my job to visit schoolchildren, both as a scientist and an elected official. I like to hear what they are thinking about and what interests them, and I love the questions they ask. My favorite classes have included making birds' nests with children, looking for snails after the rain, and writing invisible messages on mirrors.

Kindergarteners ask the best questions. Will the sun ever stop shining? Can you bring back the dinosaurs? Do you think animals can talk? I explain that the sun is a huge ball of gas in the sky and that it has been shining for a very, very long time—even before the dinosaurs; that always impresses the children. And I explain that the sun will continue to shine for a very long time to come. I share with them that "movies tell us we can bring back the dinosaurs.

But right now, that's not possible. Dinosaurs lived many millions of years ago, and today the stuff that could bring back a dinosaur only lasts for about one million years." The children are satisfied with "stuff." There is no need to discuss DNA, proteins, and soft tissue.

I tell the kindergarteners about Koko the gorilla, who used sign language, understood 2,000 English words, and had kittens as pets. The children are delightfully surprised that a gorilla could be so loving and tender. I share with them that many animals like to have conversations just like the one I am having with the children. "In a conversation," I say, "You and I take turns speaking. I speak, then you speak. Well, guess what? Many animals do the very same thing; they take turns: birds, elephants, and even whales. Can you believe it?" The children immediately understand and practice turn-taking behavior, which underpins human communication and is universal across cultures.

My experience with schoolchildren is that there are no gender gaps in curiosity or performance at this level, although there are gender gaps in toy preference at 2 years of age. According to Professor Rippon, "A close look at the wide range of statistics available shows that gender gaps do not exist at kindergarten level, but only start to appear in 6–7-year-olds and then get larger."[93] The change is not due to the emergence of a new skill but rather appears to be linked to strong external and internal views about gender and intelligence, and ultimately, about who can or cannot do science.[94]

By age six, I, too, had strong views on gender and what was or was not possible despite having two parents who could not have been more focused on equality—equal expectations of my brother and me, equal love for us both, equal division of household chores, and equal social rules. Regardless, I still grew up in a world of he, him, and his. The English language stressed "man," and masculine nouns and pronouns were used to refer to individuals and connections between jobs and gender (e.g., fireman, policeman, postman, and salesman). What my own language told me was what wasn't for me. By age six, I thought the world was an unfair place and, sadly, even kept a running tally of all the reasons why it was better to be a boy than a girl.

While gender-neutral language has largely replaced male-dominant language, common stereotypes still associate brilliance and genius more with men than women, discouraging women's pursuit of careers in fields like science.[95] Psychologist Lin Bian, now at the University of Chicago looked at how stereotypes affect children's attitudes and interests.[96] At age five, children did not differentiate between girls and boys in expectations of being "really, really smart." But by age six, girls were less likely to believe that members of their gender were bright and began to avoid activities for "smart" children—brains guide us away from things that make us feel bad.[97]

I find these results heartbreaking. Research shows that girls are equally or more likely than boys to achieve minimum proficiency levels in math and science at the upper primary level. Country assessments tell us this holds true in "54 out of 86 countries in math and 58 out of 62 countries in science."[98] In 2018, O'Dea et al.[99] undertook a broad analysis of more than 200 different studies from around the world, covering 1.6 million elementary, high school, and university students. In line with other studies, they found strong evidence for a low variation in grades between girls and boys and higher average grades for girls. They also found that the top 10 percent of a STEM class had equal numbers of girls and boys; girls were over-represented in non-STEM subjects. These tests and subsequent results confirm that girls often score equal to or higher than boys in science and math.[100] We should expect results of this nature: girls have the same biological design as boys and therefore the same potential. Nevertheless, harmful social perceptions and expectations persist, and the belief that boys are better at math and science is perpetuated by parents, teachers, media, and algorithms.

Unfortunately, the research also shows that gendered ideas about brilliance are acquired early and have a damaging and immediate impact on children's interests.[101] Each child should know that she, he, or they are special, and the world should be unfolding and revealing its secrets; instead, half of the children appear to be learning they are lesser, the world is getting smaller, and the opportunities are fewer.

These results convinced me that it is important to visit grade 2 classes. Children not only need to have scientists as role models but also need to know that they can all aspire to be researchers or work in scientific fields. Ask a grade 2 class who loves science and every hand goes up. Ask the children what science is and it's amazing what they have already picked up. People who care for animals, people who have telescopes and look at the sky. Doctors who care for babies, astronauts who go to the moon, and "big, big books!" Although they are not sure of the name, they describe, with great excitement, "The machine that lets you look at bugs in your blood." The translation of this is, of course, a microscope that allows us to see tiny microorganisms, invisible to the naked eye, and that can make us sick.

But ask children who does science, and the answer is largely men—"men in white coats." At age six or seven, they have already developed the view that it is men who do science, confirming the stereotype of the "scientist in the white coat." Where do they get these ideas? A review of online primary school science education materials revealed that 75 percent of adults pictured in science professions were men.[102] And research found that 70 percent of 500,000 people across thirty-four countries associated science more with males than with females.[103] We are telling girls that science is male when they should be learning that women work on the International Space Station, explore the depths of the ocean, and research how to protect against, treat, and cure COVID-19.

I like to let the grade 2 students in on the secret that they have been discovering, exploring, and asking questions just like scientists for years. I want them to see themselves as potential astronauts, doctors, or researchers like primatologist, conservationist, and ecologist Jane Goodall. Children love to learn about Goodall and the fact that, as a child, she dreamed of living in Africa among the animals. I explain that she had no training and no experience, but she had something very special inside her and that something is in all children. They wait on the edge of their seats for the something. I explain Jane had curiosity, like all children. She was also brave and determined, and she was willing to try something that no one else before her had. She would leave her home in England, travel to Africa, and live among chimpanzees. I explain that chimpanzees are among our very closest relatives and that it wasn't always easy for her. It was hot, there were thunderstorms, and she was lonely. One of the boys puts up his hand and wants to know if Jane forgot to bring her teddy bear. Smiling, I explain that almost immediately, Jane made amazing discoveries: chimpanzees made tools and they used them.

A little girl asked me a question: "Do chimpanzees use hammers? I use hammers and nails with my Daddy. You have to be safe when you hammer."

"That's a very good question," I answered. "And yes, chimpanzees hammer; they use hammer tools to crack open nuts. And guess what? They fish; they fish for the food they eat! Chimpanzees take a branch, they take off all the leaves, and they put the branch into the ground to look for insects."

The kids were enthralled with Jane's story. I asked them, "Is Jane a scientist?"

They all agree that she is. I then ask the children in the class, "If everyone here loves science and Jane is a scientist, why do only men do science?"

And without missing a beat, a bright little girl will say, "I can do science too; I can be a scientist. Girls can be anything we want."

While that little girl gives me hope, girls will continue to receive messages that undermine them throughout their school experience. Six-year-old girls are more likely to avoid games meant for "really, really smart" children,[104] and by grade 4, girls have less confidence than boys in their math abilities in about 60 percent of countries.[105] Research by psychologist Melanie Steffens et al. found that nine-year-old girls associated math more with males, less with themselves, and had stronger intentions to drop math, despite there being no sex difference in children's grades.[106] While girls clearly have the brains, abilities, curiosity, and building blocks of science, they are not hearing, feeling, seeing, and believing this.[107] That means it's up to parents, carers, teachers, and media to change the messages girls are receiving and to foster their natural-borne curiosity through elementary school, high school, and beyond.

* * *

We should support and encourage children's rare capabilities. They are built to explore the world and seek new experiences and knowledge. Children are focused on discovery, and we should see this wonderful characteristic in all children—girls, boys, non-binary, and gender-diverse children. Instead of seeing children as imperfect or miniature adults, as historically has been the case, Professor Gopnik sees children as "brilliant butterflies flitting around the garden exploring" and adults as caterpillars inching their way along an adult narrow path.[108] Parents, carers, and teachers all have an important role in fostering curiosity and exploration, and they must provide the right balance of nurture and autonomy because there is a real difference between the attentive adult who supports discovery and the overprotective one that stifles it.[109]

Over years of teaching dance and sport, I have learned that everyone dreams of something. If we show kids what is possible, give them space, and provide the right progressions, they cannot wait to try. If we make each new step exciting, support each of their efforts with positive reinforcement, and slide in the correction, it's amazing to watch children transform. In teaching children to dive off a five-, seven-, or ten-meter diving platform, I have never told them that the platform is not for them or spoken about how hard or scary it is. Rather, I show them the possibility by having them watch older divers and introducing them to their inspirational heroes. Girls and boys alike then push to take their first jumps off one-meter and then three-meter boards.

As in sport, success in science is built on encouragement, motivation, and achievement. Yet when it comes to math and science, consciously or not, the message we have been sending girls before most of them have the chance of entering a science lab is that this discipline is not for them. This rather toxic message is gradually changing to an acknowledgment that women and girls are missing in science and, ultimately, should lead to the more hopeful insights that the world needs science and that science needs women. Personally, I think that we must go even further, by letting girls know that we need their smarts and their ideas and that we cannot wait to see how they will change the world.

* * *

A 2020 UN Development Report showed that despite the progress that has occurred in closing the equality gap, people's beliefs still negatively impact women. The Gender Social Norms Index measured how social beliefs affect gender equality. The index was based on data from seventy-five countries that comprised 80 percent of the world's population. Over 90 percent of men and 86 percent of women showed at least one clear bias against gender equality in areas such as economics, education, intimate partner violence, politics, and women's reproductive rights. In fact, the index showed that nearly half of the world's population believe that men make better political leaders, more than 40 percent believe that men make better business executives, and bias against gender equality is on the rise.[110]

Research in language processing also shows this bias toward men. Cognitive scientists examined over 630 billion words written by millions of people on the internet to look at the concepts of person and people. What they found is that the collective concept of person is more similar to men than it is to women. While women and men each make up 50 percent of our species, people are conflated with men indicating that "the collective concept people privileges men over women." This not only has cognitive impacts but also affects society and policy-making.[111]

STEM fields have long been controlled by men and continue to be directed by them. For decades, scientists, professors, medical professionals, teachers, and the public believed that men had better brains and more aptitude for math and science.[112] In fact, in 2005, the former president of Harvard University, Larry Summers, wrongfully claimed that men outperform women in math and science because of biological differences and that discrimination is no longer a career barrier for women in academics.[113] It was wrong, ignorant, and out of touch with the realities that women academics faced. Summers resigned in 2006.

That harmful, decades-old thinking persists today. Even though the proportion of women in science courses and careers has dramatically increased in some countries over the last forty years, research from 2015 with about 350,000 participants across sixty-six countries shows that, regardless of location, the stereotype that "science is for men" is alive and well. This holds true for both explicit beliefs and implicit associations.[114] Moreover, these stereotypes are being passed on to children.[115] Over the last fifty years, almost eighty studies of the "Draw-a-Scientist-Test," in which more than 20,000 children have been asked to do what the title says, have been conducted.[116] In the mid-1960s to the mid-1970s, less than 1 percent of participants drew a woman. Today, 28 percent of drawings depict women.[117]

This change may be attributed to an increase in women's representation in science since the 1960s. However, their visibility in scientific roles is still far from parity and children tend to accept as fact what they experience in their daily lives.[118] They pick up on social cues and develop rules quickly. They accept the stereotypical division of the world into pink and blue and adapt to the idea that they ought to choose Barbie over building blocks. These gendered rules have ensured that the gap between girls and boys remains. By the time students reach middle school, more than twice as many boys as girls want to work in science or engineering-related jobs.[119]

It is time to shatter damaging myths and for science to trump folklore that holds girls back and creates self-fulfilling prophecies. To truly move science forward, we must erase the damaging stereotype that science is for men. It hurts and undercuts girls and women. Science is not a club, and least of all a boys' club. It is for everyone. The question then becomes how we best support girls who are born to do science and foster their innate curiosity and creativity.

* * *

Girls not only suffer from stereotypes and lack of role models from an early age, but they also face bias from multiple sources—some very close to home. Parents still tend to think their sons are brighter than their daughters and are two-and-a-half times more likely to do a Google search for "Is my son gifted?" than for "Is my daughter gifted?"[120] In the United Kingdom, parents and carers of boys were more likely to discuss STEM jobs with their child (70 percent) than parents or carers of girls (56 percent).[121] In Chile, Hungary, and Portugal, 50 percent of parents expected their sons to have a STEM career, but less than 20 percent had the same expectation for their daughters.[122] However, a Microsoft study in 2018 found that an encouraging mother, father, teacher, or the cumulative support of all adults could increase girls' interest in STEM subjects by 20 percent, 17 percent, 21 percent, and 32 percent, respectively.

There is much that parents can do: make science and math fun, part of play, and part of experiencing new things and exploring new places. A walk in the forest offers so many opportunities for girls to learn not only about what animals, birds, and insects live among the litter, plants, and trees, but also about how they communicate. Bees dance and chatter. They quack, toot, and whoop. Bats call each other names, use different dialects, and even talk to their young in "motherese," just like human mothers with their babies.[123] A walk in the forest offers the opportunity to talk about how woods help fight climate change and how we are all connected to nature.

In 2017, Accenture surveyed 8,500 young people, parents, and teachers across the United Kingdom and Ireland. Over 50 percent of parents admitted to having made subconscious gender stereotypes in relation to STEM.[124] It is important to pay equal attention to daughters, sons, non-binary, and gender-diverse children when it comes to science and math: studies show that families spend more time with young boys in play that involves spatial cognition and when it comes to expectations of children's performance in class.[125]

Toys offer training opportunities. If girls are not given construction toys, they may not be developing the spatial skills that will be important later in life, and if boys do not have an opportunity to play with dolls, they may miss out on nurturing skills.[126] While girls are often encouraged to play with boys' toys, we do not often see the reverse.[127] Research has shown that very young children are aware of their parents' likely approval rating of their toy choice and what is expected of them, despite what parents may think. The brain is most plastic up until about seven years of age. These first years are a critical period of development, and the more we can open the world to children and ensure equal opportunities, the better.[128]

When I was growing up, my father told my brother and me every day that he loved us equally and that he expected the same of us in school, in sport, and in life. As a child, I wondered why he repeated this mantra over and over. As I grew up, I came to understand. When teachers, coaches, or professors said, "girls can't" or "women don't," I knew they were wrong and set out to prove them wrong.

Parents and carers might also share bedtime stories about inspiring girls and women. Instead of reading children *Cinderella*, we should tell them about Caroline Herschel (1750–1848) and show them the marvels of the night sky. Although Caroline was sick as a child and her mother opposed her education and preferred that she help manage the household, Caroline became a brilliant astronomer and the first woman to discover a comet. And Caroline didn't find just one comet; she found eight.[129]

Girls find it "super cool" that comets are cosmic snowballs that orbit the sun, can be as big as a town, and have tails that stretch for millions of miles. Once, when I shared the story with a group of girls, I asked the girls what they thought. One girl put up her hand and said, "I like her [Caroline]. She didn't wait for Prince Charming. She did things all on her own." Her proud mother later explained that they had taught her that girls could forge their own paths and that they do not need rescuing. She further explained that girls who are pioneers and trailblazers have no role models and that she was raising a leader, not a follower.

The point is: watch language. Babies and children are "tiny social sponges" with a "particular appetite for social rules," and their environments and experiences clearly alter the way their brains develop.[130] Girls will pick up on cues about whether or not mom likes science and math and whether or not dad thinks they are subjects for girls and spends equal time doing these subjects with his daughter, son, or gender-diverse child. Night after night, my mom did homework with my brother and me. We were fortunate that our parents were home to do this with us—it is not possible for all families, some of whom work two and three jobs just to put food on the table. That's why it is so important for governments to fund homework and STEM clubs in communities.

In elementary school, my mom drilled into us times tables, long division, fractions, and famous train problems that required finding out when two trains heading toward each other crossed paths. She not only read through science experiments with us but also had us do them and even demonstrated hard concepts. One spring she dragged up the Christmas lights from the basement to explain the difference between parallel and series circuits.

She also demanded more of us than was expected in school. I remember proudly showing her my grade 5 project on the Siberian tiger that I had spent hours researching, sketching, and writing about. After reading it, my mom asked, "Did you think to call the museum or zoo to ask them questions about the tiger?" I had not. My disappointed mom strongly suggested that I prepare questions, call the zoo, and ask to speak to a zookeeper, all of which I did. Unfortunately, what should have been a great learning experience was undermined by my over-hearing the person on the other end of the telephone line. He did not want to waste his time and speak to a "little girl." The person at the zoo missed an opportunity to turn a child on to science.

What my mom did not tell me until the last years of her life was that she never liked math and science. When I found out, I was shocked. My mom had been a physical education teacher and had been showing my brother and me anatomy textbooks before we even went to kindergarten. She wanted us to be interested in science and did not want to pass on her insecurities to us. It is also important for parents to correct misinformation, misperceptions, and stereotypes and help children understand that all children are equal and that they should have equal opportunities.

* * *

Despite all the talk about encouraging girls in math and science, many teachers still harbor unconscious biases that discourage girls from going into these fields. In Latin America, "between 8 percent and 20 percent of grade 6 math teachers" believed that math is easier for boys to learn. Research shows that girls ask fewer questions, have less instruction and discussion time, and receive less praise than boys.[131] According to researchers Dr. Edith Sand and Professor Victor Lavy, the unconscious biases of elementary school teachers dramatically affect female academic choices later in life.[132] After analyzing math test scores of 3,000 students in Tel Aviv, researchers found that classroom teachers gave girls lower grades in math than external teachers did, while boys received higher grades. The external teachers did not know the gender of the students, suggesting that the classroom teachers were biased. The researchers concluded that in math and science, teachers underestimated the girls' abilities and overestimated the boys' skills and that this had long-term consequences for students' attitudes toward these subjects: namely, it discouraged girls from pursuing advanced courses in math and science. And it had the opposite effect on boys. Because enrollment in advanced courses in math and science in high school is a prerequisite for postsecondary schooling in computer science, engineering, and other fields, these results suggest that "teachers' biased behavior at early stages of schooling," when girls were outperforming boys, had long-term impacts on occupational choices and earnings at adulthood.[133]

The results of that study are discouraging, as are the findings of a 2018 Accenture survey of 5,000 young people, parents, and teachers in the United Kingdom. It revealed that two-thirds of teachers admitted to stereotypes about girls and boys in relation to STEM jobs.

Girls can also face structural bias in curriculum development and design. In primary schools in India, for example, more than 50 percent of illustrations in math and science textbooks depict boys, while only 6 percent depict girls.[134] While there are many challenges, boys and girls around the world enroll and complete primary school at the same rates: 90 percent of boys and 88 percent of girls enroll, and 90 percent of boys and 89 percent of girls complete primary school. This trend of equal enrollment and completion continues from primary school to secondary school. This represents a significant opportunity for further progress. Moreover, girls often do as well as, or better than, boys in science and mathematics.[135]

In traveling across the country and around the world, I have heard repeatedly from women that they remember the feeling of being ignored in elementary school and high school because they were girls and young women, and how galling it was when their test scores and grades were higher than their male colleagues. Worse, the better experiences and opportunities were offered to the boys and young men.

Teacher training must include the latest research that shows that brains are plastic and girls have the same abilities as boys. Teachers should have the same expectations of girls, boys, non-binary, and gender-diverse children and ensure that all children have equal experiences and opportunities. More work needs to be done to build equal, inclusive, welcoming classrooms.

If we want more girls interested in math and science, there needs to be an inclusive curriculum, discussions, and posters featuring inspiring science heroines, like two-time Nobel laureate Marie Curie, or explorer, pioneer, and icon Dr. Jane Goodall. Additionally, opportunities for children to engage in citizen science projects should be provided. What I remember throughout elementary and secondary school were periodic tables and pictures of the great male scientists stretching the length of each classroom wall (just as iconic male musicians covered each orchestra classroom). Textbooks must include the mothers and fathers of science, their science journeys, and show girls and women succeeding in science and STEM fields—not just as anecdotal sidebars in the history of science. And teachers should invite scientists of all genders into the classroom so that children have the opportunity to meet them; get to know them as individuals; ask questions about her, his, or their journey to becoming a scientist; learn about their work; and, most importantly, get a taste of real science.

* * *

Elementary schoolchildren are most often exposed to their parents, teachers, and media, so what parents, teachers, television, and the internet say matters. While media offer channels of communication that can reach millions of people, research shows that parents, teachers, and the media present stereotypes of who does science and what scientists look like. A 2018 study by the Lyda Hill Foundation and the Geena Davis Institute for Gender in the Media shows that media often sends the negative message that STEM is for white men. The media has the power to educate and inform, to challenge, engage, and inspire. It should reflect and represent diverse communities.

Female characters like Dana Scully (*The X-Files*), Meredith Grey (*Grey's Anatomy*), and Amy Farrah Fowler (*The Big Bang Theory*) have inspired girls and women to pursue STEM.[136] We know that girls and women are more likely to go into STEM if they personally know someone in STEM, have a STEM role model, and have family members, friends, and teachers who encourage them to pursue STEM.

While there are innumerable stories to tell about girls and women making cutting-edge discoveries, these stories are rarely brought to popular attention. That's a missed opportunity to change how girls and women think about themselves and their futures. The 2018 report, "Portray Her: Representations of Women STEM Characters in Media," showed that the entertainment media has a long way to go to improve stereotypes about pursuing STEM careers.[137] It showed that male STEM professionals portrayed in the media outnumbered women by nearly two-to-one and had not improved in a decade.[138] The vast majority of STEM characters, disproportionately to the general population, are white (71 percent), with fewer being Black (17 percent), Asian/Asian-American (6 percent), Latinx (4 percent), or Middle Eastern (2 percent).

Indigenous girls, girls from racialized backgrounds, girls with disabilities, and girls who belong to the 2SLGBTQI+ community need to be represented. They need to see that they belong and what is possible for them. Girls are not one-dimensional. Rather, they are a beautiful intersection of gender identity, nationality, race, disability, sexual orientation, and more. Girls are Black, newcomers, and refugees; they have disabilities and are gay; and they need to know that they matter and that they have a place.

We hear the importance of increasing the number of girls and women in STEM from the United Nations, and we hear it from our governments. There is a push to do better, and girls feel that push. But if the message is undermined, even unwittingly, by society, parents, teachers, and the media, girls will pick up on it. If the media always goes to an older white male scientist as the "expert," this, too, will have an impact, as will caricatures of women scientists with white coats and glasses.

Recently with COVID-19, broadcasters clearly tried to present a balance of scientists, showcasing not only female and male but also highlighting racial diversity—a welcome step. It is my absolute hope that all girls know that they belong, they matter, there is something special in each and every one of them, they have the opportunity to discover and explore every day, and they are encouraged to reach for the stars. I also hope that, beyond a loving parent or carer, each girl has a teacher or coach who, when the work seems impossible or overwhelming, will step up, support her, challenge her, and remind her that the opportunities are endless—that she will learn things and do things that we cannot even imagine.

CHAPTER 3

Socialization, Social Brain, and Science

FOR MILLENNIA, WOMEN'S LIVES were regulated by men. Fathers, husbands, brothers, grandfathers, uncles, and teachers have together created a male-dominated world and culture. During the Renaissance, economics, politics, education, religion, and sexuality were systematically controlled by patriarchal social structures. "An obsession with protecting the purity of the male bloodline could make women virtual prisoners in their own homes."[139] In mid-eighteenth-century England, women who were interested in intellectual pursuits were often referred to as "blue-stockings," a derisive reference to an informal social and educational movement composed of women who dared to aspire to learn. These courageous women were seen as both off-putting and unfeminine.[140] In Victorian times, gender roles became even more sharply defined. Women and men inhabited "separate spheres": men worked, while their crinoline, morally superior, physically weaker wives, daughters, and sisters stayed home and oversaw domestic duties.

While it was acceptable for women to be educated in certain cultural "accomplishments" such as dancing, drawing, modern languages, music, and singing, doctors warned that too much education could hurt their ability to reproduce.[141] For hundreds of years, male-generated science tried to explain why women were intellectually inferior. Even history's most famous biologist, Charles Darwin, believed that women lacked male capabilities. He failed to consider the cultural and social norms that controlled women.[142]

This chapter tackles socialization and the social brain—the network of brain regions that are involved in understanding others. It examines how socialization influences a developing brain, how it has impacted women and science throughout history, and how it continues to hold girls and women back today. It also identifies the social foundations of power structures that continue to impose barriers and obstacles that girls must navigate and overcome to succeed in science. We already know that girls have the right brain for science. They also have the right building blocks—brilliance, personality traits, and systems thinking—to advance science as a discipline and to transform our societies.

Socialization is the process that introduces people to interpersonal norms and customs that allow them to function in society and, in turn, allow society to operate. Families, schools, peers, and media all play important roles in a person's socialization. Through socialization, people learn to become members of a group or community, as they align their values with specialized social norms to prepare themselves for different roles.[143] Culture, history, language, institutions, race, age, economic circumstances, and gender all contribute to a person's socialization.[144] A 2018 study by Neubauer et al. showed that 300,000 years ago, the brain size in early *Homo sapiens* "already fell within the range of present-day humans," and the brain shape did not reach the present-day variation until between 100,000 and 35,000 years ago.[145] A 2022 study by Villmoare et al. found that the human brain size has not changed in 30,000 years and probably not in 300,000 years.[146] While the brain has remained the same size over the last millennium, modern society has encountered significant changes, leading to a populace with increasingly varied socialization.

Regressive and repressive mechanisms of existence, like colonialism, slavery, and gender inequality have likely existed for millennia. Humans have lived in a male-dominated world and culture where gender inequality was and remains the norm—a reality that hurts everyone. Men set the standards and wrote the rules, largely to the benefit of men.[147] As they worked to maintain their power and control, women's roles changed in response to men's shifting fancies and needs: from the Renaissance through to Victorian times and the Second World War, when women drove trucks, operated heavy machinery, and worked in munitions factories, and on to today. These changing roles have not served women and men equally.[148]

I present the following information with the recognition that women and men are not uniform groups, that not all people can be lumped together under these terms, and that social norms vary across geographic regions and time. For example, in seventeenth-century and eighteenth-century Europe, women were expected to be curvy or have a rubenesque figure. Corsets held in waists and supported bosoms, and the shape of the corset changed with shifting societal views. Women's lower halves were caged, hooped, or padded to create volume. The early part of the twentieth century brought a shift in preferences for athletic, slender, young bodies, along with an epidemic of eating disorders.[149]

For centuries under Anglo-American common law, a woman or girl was passed from her father's authority to her husband's. A husband exercised near-exclusive power and responsibility over his wife. He was entitled to sex and rarely had to consult his wife to make decisions about property matters, although women found ways to navigate "the structures of authority that governed them", and the law increasingly undermined patriarchal relationships.[150] Marital rape was not criminalized across all fifty U.S. states until the 1990s.[151]

Hundreds of years ago, women were more likely to be seen as experts in superstition and witchcraft than in science. Witch-hunts swept across Europe in the sixteenth and seventeenth centuries; the vast majority of those prosecuted and executed were women.[152] Although women were thought to have had the wrong brain for science, many pursued scientific disciplines in those days, provided they were able to gain access to education or had the means to do so. They became specialists in disciplines such as astronomy, botany, or geology. However, as science evolved from being an unregulated pursuit to an institutionalized profession, women began to be deliberately excluded.[153] The Royal Society, a fellowship of many of the world's most eminent scientists and the oldest scientific academy in continuous existence, was founded in 1660, but the first women fellows were not elected until 1945.[154] Many women scientists were simply written out of history,[155] although researchers such as Margaret Rossiter and Londa Schiebinger have been working to bring to light "the ingenious accomplishments" of those who have been forgotten.[156]

In the early nineteenth century, a struggle over nomenclature emerged. No one really knew what to call someone who studied science professionally; discussions involved the names "cultivators of science," "men of learning," "natural philosophers," "savant," and "men of science." It was not until 1834 that Cambridge

University historian and philosopher William Whewell came up with the term *scientist*, a linguistic analog to the artist and a term that could bring unity to the various branches of the sciences.[157] While the term was quickly picked up in the United States, "man of science" was preferred in Britain.[158] "Man of science" was not only consistent with "man of letters" but it also commanded respect and showed that science was pursued by the more intelligent, serious sex.[159] In 1924, physicist Norman Campbell sent a letter to the editor of *Nature* and asked him to reconsider the journal's policy of avoiding the word "scientist." He argued that "a shibboleth was no longer needed," and that the continued use of "man of science" was offensive to feminists.[160]

While today we take for granted that women can choose education, own property, and be a scientist, we must remember that these are hard-won rights and that antiquated understandings of women continue to influence our thinking today. It is important to keep in mind that our understanding of any particular social practice may be based on biased information and supported by the same guidelines that have held women back for generations.

* * *

It is important to remind ourselves in any discussion of socialization and science that boys and girls start on a level playing field in terms of ability. Their academic achievement and academic performance are similar. In 2015, the Trends in International Mathematics and Science Study (TIMSS) compared the performance of grade 8 students across thirty-nine countries in 2015 and found no gender difference in STEM achievement at that level. That same year, the Programme for International Student Assessment (PISA) looked at fifteen-year-old students in seventy participating countries and found that in about one in three participating countries, there was no gender difference in science achievement; in the remaining countries, the gender gap was shared almost equally, either in favor of boys (34 percent) or girls (31 percent).[161]

The real difference between girls and boys may be attributed to the environment in which they grow up.[162] There is a real opportunity for parents to help their children transcend narrow, gender-driven expectations. The first requirement is an awareness that gender expectations and roles affect nearly every aspect of life from infancy forward.[163] As we learned in the previous chapter, they are communicated through such social conventions as color-coded toys and types of play.[164] Stereotypically, girls receive dolls and dollhouses, while boys receive cars, Lego, and toy soldiers. If given a choice, children may well express entirely different individual preferences. As girls get older, social norms dictate that their toys emphasize beauty and nurturing, whereas boys' toys tend to stress action and excitement.[165] Toys affect not only how children see themselves but also the skills they develop and their brain development.[166]

Parents socialize gender in many ways: they are role models and they reinforce and shape children's behaviors, select children's environments, and provide opportunities.[167] A 2017 Pew Research Survey showed that "76 percent of Americans found it to be somewhat good or very good that parents steer girls toward boy-oriented activities and toys." In contrast, only 64 percent felt parents should encourage boys to participate in activities and play with toys usually associated with girls. The asymmetry in these findings held up across genders, generations, and political groups.[168] Traits stereotypically associated with boys and men continue to be seen as good and something to be admired, while traits stereotypically associated with girls and women are considered to be lesser and something to avoid.[169]

School is another important source of socialization for students of all ages. In class, young people experience spoken, written, and unwritten rules and expectations related to their behavior, schedules, and tasks. Their conduct is either reinforced, rewarded, or penalized. Teachers model behavior, provide expectations, offer opportunities, and provide evaluations; they can also provide stereotypic educational materials. Some teachers attribute intellectual talent to boys, even in the case of failure—boys fail because they do not use their full capacity. While boys become more confident—often beyond their capabilities—girls are more likely to underestimate their abilities, especially in areas like math and science. Peers also reinforce gender roles through their own gendered behavior, non-verbal behavior, and punishing comments for those who do not meet their gender expectations. By the time they reach high school, girls can internalize these messages, and for a while, it may have seemed cool to some to wear mass-produced t-shirts with the messages, "Allergic to Algebra," and "I'm too pretty to do homework, so my brother has to do it for me."[170]

The 2017 Common Sense Media report, *Watching Gender: How Stereotypes in Movies and on TV Impact Kids' Development*, showed that constant exposure to the same dated concepts in the media, starting before preschool, stuck because they were targeted at children when they were most receptive to their influence.[171]

In movies or on TV, girls are often shown as overly interested in fashion, their looks, and as over-sexualized and eager to win over a boy. Males are often portrayed as capable problem solvers. In fact, whenever we watch a movie, chances are high that we are watching something from a male perspective: in 2021, "94 percent of the top 250 grossing films had no women cinematographers, 92 percent had no women composers, 82 percent had no women directors, 73 percent had no female editors, and 72 percent had no female writers." Of the top 250 grossing films, 61 percent included four or less roles for women, 32 percent included five to nine women, and just 8 percent featured ten or more women. In stark contrast, only 4 percent of films included four or fewer male roles and 72 percent had ten or more men.[172]

The film industry is no different than many other power structures that affect our lives. Male-dominated social structures broadly underpin corporations, economies, political, and science systems. They entrench and perpetuate women's subordination in the boardroom, parliament, and science laboratory. According to the 2021 World Economic Forum report, it is expected to take over 250 years to close the economic participation and opportunity gap. If the average time span between generations is twenty-five to thirty-five years, that generally means there are three to four generations every one hundred years. Assuming a similar rate of generational change over the coming 250 years, that will mean eight to ten more generations of economic inequality if current social structures and the current rate of progress persist.[173] Meanwhile, a 2016 survey found that there were more CEOs of large U.S. companies named John than there were CEOs who were women.[174]

Human beings are innately social. We need social connection more than anything else to survive because we are effectively born incapable of looking after ourselves.[175] "Neuroimaging and lesion studies have identified a network of brain regions that support social interaction;" that network is often referred to as the "social brain." Each region in the social brain likely contributes to a specific type of social processing.[176] For example, the almond-shaped amygdala, which is found in both the right and left hemispheres, is important for the interpretation of emotional facial expressions and for tagging group membership, like parents and caregivers, as well as for coding out-groups.[177] Neuroimaging shows that although newborn infants have all the right brain parts to support social interaction, these regions are not yet connected or specialized in the right way. This means that much of our social ability develops and improves over time with repeated social interactions and continues to specialize into adulthood.[178] Over time, our sense of self, sense of others, and sense of belonging develop, and we learn social norms and social rules.[179]

Our social brain is the basis for our evolutionary success: it allows us to understand others, to collaborate, and to cooperate. Because group membership has been essential to our progress as a species, we are good at determining who is part of our group and whether or not we fit. In fact, our self-esteem is influenced by how well we are rooted in our social groups.[180] The structure of the social brain develops substantially throughout adolescence. Adolescence is marked by heightened sensitivity to social stimuli, the increased need for peer interaction, and the importance of peer acceptance and peer influence.[181] "Multiple longitudinal MRI studies have shown that the volume of grey matter—mostly consisting of cell bodies and synapses—declines from late childhood to the mid-twenties, whereas the volume of white matter, consisting of myelinated axons, gradually increases."[182] Professor Matthew Lieberman explained in his TEDxStLouis talk that on the outer surface of the brain, there is a network for analytical thinking while the network for social thinking is located along the midline of the brain. This social network in the brain tends to be tamped down during other kinds of thinking but "pops up like a reflex" in any spare moment so that we can see and experience the world socially.

Our brain not only processes information but looks for predictability and generates rules to guide us safely through the world.[183] It is highly sensitive to social context—the framework in which information is presented. Experiments show that the same task given with positive context can lead to the correct part of the brain being engaged and better performance, whereas negative context can lead to brain engagement associated with mistakes and poor performance. The human brain constantly monitors what is going on around us, learning and relearning the rules of social engagement which begin early in life. It allows us to change our behavior to fit into groups and to develop a self-identity that fits in with the people around us.[184] We learn that different behaviors are expected in a church, a gurdwara, or a masjid than at a soccer match, the cabinet table, or the supper table.[185]

Avoidance of pain is a powerful motivating force and drives human beings to extraordinary lengths to avoid hurt or do the right thing to be socially accepted.[186] Social psychologists Matthew Lieberman and Naomi Eisenberg set out to test whether social pain, the experience of pain because of interpersonal loss or rejection, is more than a metaphor. Their experiments involved people lying in an MRI machine where they believed they were playing a simple, online, ball-tossing game with two other people in scanners. But part way through, the other people stopped throwing the ball. The same brain regions that register distress with physical pain were activated when people were left out of the game. The worse people reported feeling about being left out, the greater the response in the regions (and, yes, Tylenol made these effects go away). In short, being bullied, bored, or excluded can hurt as much as an electric shock. Negative experiences can impact self-criticism, self-esteem, and self-silencing.[187] More recent studies have repeatedly shown that social pain activates the same networks as physical pain and that it can upset us for days, weeks, and months.[188]

* * *

The understanding of what science entails has changed significantly over time. Until the nineteenth century, it was understood that it included natural philosophy in conjunction with other fields including astronomy, chemistry, mathematics, and physics.[189] It was only in 2009 that Britain's Science Council published what it claimed was the first official definition of science, namely that "science is the pursuit and application of knowledge and understanding of the natural and social world following a systematic methodology based on evidence."[190] While this short and simple definition is meant to capture a wide range of activities, it does not mention that science is also a body of knowledge, nor does it encompass the idea that science is socially embedded and socially influenced. Science progresses by asking questions, running experiments, collecting data, and asking new questions. Those questions, in turn, are influenced by social and cultural context: who does science; how are experiments conceptualized and conducted; who is selected for experiments; how are data reviewed, interpreted, and reported; and how are new theories developed.[191]

In 2016, the British Science Council added that a scientist is "someone who systematically gathers and uses research and evidence, to make hypotheses and test them, to gain and share understanding and knowledge."[192] If a scientist is "someone" who uses research, how many of the legendary women scientists can

SOCIALIZATION, SOCIAL BRAIN, AND SCIENCE

you name? This is something I have always asked my university students. "I will make it easy," I tell them, "I will take any woman in a STEM field." The first women named are usually the same: Marie Curie, Ada Lovelace, Jane Goodall, and Katherine Johnson. Beyond that point, the students often struggle to find another name. I then ask them to name famous male scientists: from Galileo Galilei through to Stephen Hawking, the list is long and detailed and, as one student usually points out, these were the men on the walls of their science classroom and in their textbooks.

Looking at the history of science is an essential part of science. It helps us understand how science developed, how science is done today, and the biases that have taken hold and prevent it from being aligned with its principal aim—human discovery. It tells us, for instance, that in Victorian Britain, the idea that women with their lesser brains might be capable of science—save perhaps botany and geology—was ridiculous. In 1860, Thomas Huxley, best known as "Darwin's bulldog," wrote privately to his friend, geologist Charles Lyell: "Five-sixths of women will stop in the doll stage of evolution, to be the stronghold of parsonism, the drag on civilization, the degradation of every important pursuit in which they mix themselves—intrigues in politics and *friponnes* in science."[193] He also espoused a racial hierarchy of intelligence and is prompting a long overdue discussion of science's contribution to racism, sexism, colonialism, and how to decolonize.

Despite Huxley's rant, women were absolutely involved in science. In 1666, Margaret Cavendish, the duchess of Newcastle, wrote *Observations upon Experimental Philosophy*. She attended a meeting of the Royal Society, against the protests of some of its all-male fellows.[194] In 1834, Mary Somerville surveyed contemporary sciences, drew attention to their underlying methodology, and wrote *The Connexion of the Physical Sciences.* William Whewell recognized her as the first "scientist."[195] These examples notwithstanding, men did their very best to keep women out of established scientific organizations and associations. Women were also excluded from the Académie des Sciences in France until 1962, the American National Academy of Sciences until 1925, and the Russian National Academy until 1939.[196] The likes of Mary Somerville have been treated by history as mere curiosities amid the powerful men of science.[197]

The perception that a scientist is male and white remains today, even though science developed "over thousands of years by people from a diversity of cultural traditions".[198] A 2015 study by Miller et al., involving 350,000 participants from sixty-six countries, showed that science is still seen as a predominantly male profession, but the more women enter the field, the less people feel this way.[199] In that same year, the L'Oréal Foundation asked 5,000 people across Europe their views and perceptions about scientists: 67 percent of Europeans did not believe women had the skills to be scientists and almost 90 percent of respondents felt that women had the aptitude for "anything but science."[200] More recently, a 2019 online survey of more than 1,500 Canadians commissioned by the non-profit group Girls Who Code showed similar results. Half of Canadians could not name a woman scientist or engineer and more than 80 percent pictured a man when asked to imagine a computer scientist.[201] Without further work to correct this systemic bias, the stereotype of the male scientist does not appear to be going anywhere anytime soon.

I remember very clearly when I began to lose interest in science. It was in my grade 8 science class, when we were told that "science is for boys" and that they would "take their first steps to becoming electricians and engineers." I wondered what options there would be for me. While the set up for the class was bad, the lessons were worse. Each of us received a small coaster with wires and light bulbs. We were to create a series circuit. The science wasn't explained and most of us just enjoyed connecting the wires and seeing the bulbs light up, much like people used electricity in the 1700s for magic tricks. The first three light flashes were somewhat impressive, but I quickly lost interest. I had a hundred questions about electricity, none of which my teacher was prepared to entertain: how does it come inside my house; where does it come from; is the electricity in my house the same as the lightning in the sky; is static electricity the same as the electricity that comes out of the wall socket; and if electric eels can really make electric currents, how come they don't get shocked themselves?

While my questions were ignored, we did learn about Benjamin Franklin (1706–1790), the author, diplomat, inventor, publisher, scientist, and statesman and his famous kite experiment that proved the connection between lightning and electricity. Selective accounts of history abound throughout this era of white patriarchy. Our teacher failed to mention that, as a young man, Franklin was a slave owner, that he regularly carried advertisements for the sale of slaves in his newspaper, the *Pennsylvania Gazette*, and that it wasn't until after the ratification of the U.S. Constitution in 1788 that he became an outspoken opponent of slavery.[202] Why was this important history left out from Franklin's story? Why was there no mention of the great women who helped to power society and to open the nuclear age, scientists like Marie Curie, Lise Meitner, and Irène Joliot-Curie?

We must seize the opportunity to complete these important historical accounts. I will start with one of the women who achieved firsts in electrical engineering, a woman named Elsie MacGill. After she graduated in 1927 from the University of Toronto, she took a job at the Austin Automobile Company in the United States and earned a master's degree in aeronautical engineering (1929) from the University of Michigan. She became the first practicing Canadian woman engineer and the first woman aeronautical engineer in the world. On the eve of her 1929 graduation, she fell ill and, the next morning, woke paralyzed from the waist down. She had contracted a form of polio and was told that she would never walk again.[203] She wrote articles about aircraft, aviation, and flying from her bed. And after three years of bed rest, she eventually walked again, first with the aid of two metal canes and later with one. She recovered sufficiently to begin her doctoral work at MIT before being lured back to Canada.[204]

In 1938, Elsie MacGill became the chief aeronautical engineer of the Canadian Car and Foundry Co. Ltd. (Can Car) in Fort William, Ontario, where at age thirty-five, she headed the Canadian production of the famous Hawker Hurricane fighter planes for the British government during the Second World War.[205] The first forty planes came in ahead of schedule in time for the Battle of Britain—a game-changing air battle waged between July and October 1940 that "pitted a small group of Allied fighter pilots against a far larger German force" and that ultimately prevented Hitler from invading Britain.[206] At peak production, "Can Car employed 4,500 workers, half of them women, and produced three to four planes per week." By the end of the war, the factory had produced over 1,400 fighter aircraft.[207]

Elsie MacGill was a pioneer, war hero, social advocate, inspiration, and role model for people of all abilities, and I would have loved to have known about her. But my grade 8 teacher, who was likely handed a curriculum written and approved by a male-centric school board, failed to provide a balance to his stories about all the men who accomplished something for the very first time with women who achieved firsts in society. Women were systematically left out of science classes and science history, and every girl in my own class was told in one way or another that she did not belong in the future of science.

Human beings are primed to be social and are socialized into a gendered world with institutions, structures, and a layer of hidden inequality that are meant to meet the needs of only half the population.[208] This inequality damages everyone—women, men, girls, boys, and sexually and gender-diverse people. It also damages relationships, communities, economies, health, and the quality of the science itself.[209] While girls and women have the right brain, the right fundamentals of discovery, exploration, and questioning, they are socialized into a patriarchal world where women are worse off than men, simply because they are women, and they are repeatedly undermined—their intelligence, their personality traits, and how they think—hurting their self-esteem and confidence.

For years, we have been socialized to think that men have an increased aptitude for genius or giftedness. New research led by Nanyang Technological University, Singapore, in collaboration with New York University, involving almost 400 Chinese Singaporean parents and about 350 of their children, aged eight to twelve, shows that children hold stereotypical views that "brilliance" is a male trait and that this belief strengthens over time. This research shows that stereotypes can take root in childhood, undermining abilities, influencing life choices, and becoming a self-fulfilling prophecy.[210] Recent research has also pointed to "gender stereotypes that portray men as more brilliant or inherently talented than women," and that these likely hold women back in careers that are perceived to require "brilliance."[211]

At the University of Washington, over 1,700 biology undergraduates were asked to name classmates whom they thought were "strong in their understanding of classroom material." Male students underestimated their female peers and nominated men over better-performing women. In fact, women needed to be awarded an A to earn the same prestige as a male student earning a B. The male nominators' bias was nineteen times that of females, and male bias only got worse with time.[212]

Princeton Professor Sarah-Jane Leslie and a group of American colleagues asked 1,800 participants across thirty academic disciplines to rate their agreement with statements regarding their disciplines, such as, "Being a top scholar in my field requires a special aptitude that just can't be taught." In some fields, success was seen as a result of determination and hard work, but in others, success required "a special, unteachable, spark of brilliance." The researchers found that fewer women participated in disciplines that people perceived to require innate talent, likely because they had been the target of negative cultural stereotypes about intellectual ability.[213]

In 2022, Clotilde Napp from the Université Paris Dauphine and CNRS and Thomas Breda from CNRS and the Paris School of Economics presented 500,000 fifteen-year-old students across seventy-two countries with the sentence: "When I am failing, I am afraid that I might not have enough talent." They found that in seventy-one of seventy-two countries studied (Saudi Arabia was the outlier), girls were more inclined to attribute their failures to lack of talent than boys, even when performance was equal. Stereotypes were stronger in "more developed or gender-egalitarian countries" and "among high-achieving students," suggesting that as countries develop, the stereotypes do not disappear. The researchers also showed that there is a strong correlation between the idea of being less talented and three other indicators: self-confidence, competitiveness, and willingness to work in fields like information and communication technology.[214] "The less talented girls believe they are compared to boys, the less confidence they have, the less they enjoy competition, and the less willing they are to work in male-dominated occupations."[215]

The other side of the coin is that 79 percent of high school girls say they feel "powerful" when given the opportunity to participate in STEM activities.[216] The Microsoft study also showed that knowing a woman in STEM personally increased girls' feeling powerful doing STEM from 44 percent to 61 percent, seeing the relevancy of STEM from 53 percent to 73 percent, and knowing how to pursue STEM careers from 51 percent to 74 percent.[217]

It is long past the time to stop telling girls and young women they cannot or should not aspire to certain academic goals. We need to be mindful of the messages we are sending. We need to develop new ways to bolster the confidence of young girls, provide new experiences, and highlight their achievements. While the required foundations are innate across genders, science as a profession requires learning. It must be practiced in the same way that someone learns to play the piano or violin. Success comes through trial and error.

* * *

Historically, whatever was most prized in society, men had attributed to them, and women lacked. At times, when the soul was revered, for example, women had none.[218] When the power of reason was valued most, women's bodies, brains, and intellect were found wanting, and when scientists discovered glands and hormones, women were overlooked again.[219] Women were to be obedient wives and devoted mothers, and non-conformity to those roles was often seen as a cause or effect of a disordered body and mind throughout the history of Western medicine.[220] In the nineteenth century, clitori-dectomy, the surgical removal of the clitoris, could cure hysteric and nervous disorders, and in the twentieth century, lobotomies— notorious surgical procedures that severed connections in the brain—could keep women calm and cooperative.[221]

By this time, it will be no surprise that girls and women not only had the wrong brains, lacked genius, and lacked a "math gene," but they also had the wrong personality traits.[222] English clinical psychologist Simon Baron-Cohen put forward the empathizing–systemizing (E–S) theory on the psychological basis of autism and male–female neurological differences. This theory claims that the "male brain" is predominantly hardwired for understanding and building systems whereas the "female brain" is predominantly hardwired for empathy.[223] Baron-Cohen also argues that boys are biologically programmed to focus on objects, predisposing them to math and understanding systems, while girls are programmed to focus on people and feelings. In other words, women are supposedly more interested in people than things, and therefore do not choose STEM subjects because they fall into the "things" category.[224] (Baron-Cohen and Rippon have debated whether or not men and women have different brains. While they agree on the dangers of pseudoscience and both want a society free of discrimination, they disagree on essential differences between the sexes. This book comes down on the side of Rippon. Their debate is easily found online.) Baron-Cohen appears to have somewhat modified his views about the role of gender in his new book, *The Pattern Seekers: How Autism Drives Human Invention*.

According to Professor Elizabeth Spelke, there is ample literature that contradicts Baron-Cohen's study and provides evidence that male and female infants tend to respond equally to people and objects.[225] Yet, researchers have found that individuals in the sciences tend to possess a cognitive style that is more systemizing-driven than empathizing-driven, whereas those in the humanities possess a cognitive style that is more empathizing-driven. Within the sciences, the systemizing pattern is especially pronounced in engineering and physics.[226] A recent study of over 400 students from the humanities and the physical sciences found an interaction among sex, degree, and the drive to empathize or to systemize. "Male students in the sciences had a stronger drive to systemize than to empathize in comparison to females in the sciences."[227] Here, we see clear links being made among male brains, systemizing, and science. I think it is important to pay attention to how culture, judgment, and values shape scientists' work, and how we have plastic, flexible brains, and flexible, plastic personalities.[228]

Our personalities can fluctuate with age, change because of life events, or shift because of the things we eat. We can alter our personalities through our own efforts, yet the belief that the male brain exists, is somehow better, and primed for science and research persists in our academic institutions. Despite all this evidence, some scientists still intentionally draw sexist conclusions. They focus on any minutia they can find to support the claim that women's and men's brains are different and their abilities, genius, and intellect are different, and that any variations are more important than the considerable overlap between them, often while arguing for equality in the world.[229] It's past time to use the scientific method itself to perpetuate claims that have held women back for centuries.

From preschool through high school, most children encounter some stereotypes that can set them on a lifelong path or lead them away from another one. It can be as simple as the social messages sent to girls—"Be a good girl," reinforcing compliance—and those sent to boys—"Boys will be boys," allowing for greater freedom and discovery. These stereotypes are even more challenging for non-binary, gender-diverse children and are notoriously sticky.[230]

We all feel pressure to fit in with the groups in our lives, be they middle school girls, high school girls, or scientists, and we tend to behave accordingly, adopting the standards, beliefs, prejudices, etc., of those with whom we want to identify. Our brains absorb and reflect the attitudes and expectations of those around us. Stereotypes can change the brain and steer behavior, paying much more attention to negative aspects while building up a picture of a group.[231]

It is possible to track "brain changes associated with acquiring a stereotype" and show how the "brain responds when there is a disconnect between the expectation that has been set up" by a stereotype and reality. If our self-image includes membership in a negatively stereotyped group, it can lead to stereotype threat: that is, anxiety or concern in a situation where a person has the potential to confirm a negative stereotype about their social group. For example, if girls are told that people like them tend to perform badly on tests similar to the ones they are about to take, they start to overthink problems and check for mistakes. This can affect how their brains process information and they underperform. But if they are told girls generally perform better on these kinds of tasks, brain imaging shows more activation in task-appropriate areas of the brain and the difference disappears. This is a direct effect of stereotype lift.[232]

Social stereotypes tend to be self-sustaining through individuals or through society and because people act according to the message in the stereotype, they tend to be reinforced through time. Stereotypes also influence society's behavior toward a group that is stereotyped. "As we grow older, we don't grow away from stereotypes." Rather, they "continue to mold our brains and behaviours."[233] It is, therefore, critically important that horizons are broadened, and opportunities are visualized among children, across genders. Girls are socialized and raised in a gendered world where some parents, teachers, and the media reinforce dated stereotypes. Imagine that a girl wants to do science: is she alone in her interest or is there a robotics team or STEM club who will support her? If the answer to the second question is yes, is she merely a number to allow a club to compete at a competition—just counting girls is not enough—or is there a concerted effort to ensure that she is included, that she feels she belongs, and that her self-image is strengthened?

More than ever, I hope that girls will be confident, proud of their extraordinary abilities, and willing to stand up for themselves and others. It is important that scientists visit schools, and girls see leading women scientists as role models and know that being a scientist is possible. Girls also need to know that these jobs are creative, exciting, and that they can have a positive impact in their communities, country, and the world. Changing the image of what science is, who does science, and who belongs in science may encourage more girls to give it a try.

These changes may start in the classroom. Teachers can play a really important role by creating inclusive classrooms, sharing stories of women scientists and STEM heroes, bringing in women role models, hanging posters of women scientists and STEM leaders in the classroom, using science textbooks that not only include women, but show equal representation, and perhaps most importantly, valuing girls' questions and contributions. Today, how do we ensure that teachers teach the whole class and don't focus attention only on their "academic stars," further alienating those girls, boys, non-binary, and sexually and gender-diverse students for whom science and math may not come easy, but who might be really interested?

A lack of inclusion really hurt one of my students. Let's call her Lisa. Throughout high school, she struggled in math, failed classes, and repeated them at summer school. Eventually, the school bumped her down to basic level, and in so doing, sealed her academic future: she would not be able to attend university. I couldn't understand it. I knew she was bright, but she fell further and further behind. I worried when I began to teach her dance theory, which involved math. She "hated math." Surprisingly, she didn't just do well in theory, she excelled. I told Lisa how smart she was and she beamed. No teacher had ever told her that before. I told her mother the same. I told her to stop comparing herself to everyone else in math class and instead focus on her own performance and always go for a "personal best," just like we do in dance. I explained that if she could do fractions, geometry, and theory while counting bars and beats, she could absolutely do high school math and anything else she wanted. From that very day, Lisa had a new goal. She was going to get out of basic level, improve her grades, and be the first person in her family to finish university. I will never forget the call when she got into the program of her choice. The call was one of the most wonderful moments in our shared lives, at least until she graduated from university. Later, after being on the leadership team for her graduate program and graduating with a master's degree, she continued to do research for the university.

All children need a teacher, coach, or mentor who is going to be there on the hard days to advocate and fight for them. We need to make it clear that science and math are for everyone. All children and youth should be made to feel they belong. There is no more basic need. They should also be made to feel they can achieve, overcome challenges, ask important questions, and that through science, they can choose to make a real difference in other people's lives and in their world. Girls want to know the stories of role models, how they overcame obstacles, and what advice they have for them about life and career.

Alice Parker was a young Black woman from New Jersey who filed a patent in 1919 for natural gas-powered central heating. Cold winters and an ineffective fireplace inspired Ms. Parker to develop a much more convenient way for people of the day to receive heat: they would no longer have to leave the house to buy or chop wood or check a burning fire through the night. And while her initial design was never used, her idea was a major step toward the heating systems in use today.[234] I tell children and youth in our community about Alice Parker, because she was young and didn't wait for others to act. She saw a problem and came up with an idea to help make life better for families and communities. What makes her achievement and contribution even more remarkable is that she filed a patent before both the civil rights movement and the second wave of feminism.

Girls learn that STEM is male dominated. They hear about abuse, discrimination, and sexual harassment, and they don't want to be part of it. To create a more equal future for us all, our focus must be on fixing the workplace and giving girls the experience. It's not one or the other; it's both. Yet governments and society tend to focus largely on girls rather than the pervasive problem in the workplace that often requires action by multiple levels of government, the private sector, and institutions. Six in ten girls admit that they would feel more confident pursuing a STEM career if they knew that men and women were equally employed in these fields.[235] That means our collective failure to address equity, diversity, and inclusion (EDI) in colleges, universities, and the workplace is turning the next generation of girls away from the highest growth careers.

A 2016 survey commissioned by Microsoft of 11,500 girls across twelve European countries showed that girls tend to become interested in STEM subjects around age eleven, but significantly lose interest between the ages of fifteen and sixteen. Conformity to social expectations, gender roles, lack of role models, gender stereotypes, and perceived inequality in the workforce are taking a toll on girls and their choices.[236] This study suggests that "governments, parents, and teachers have about a four-year window to inspire girls' passions before it closes." It also means that governments and societies have a tremendous amount of work to do, work that they have collectively failed to do for a century or more, to achieve equality in STEM fields.

To have more girls and young women choosing to take advanced STEM classes at the end of high school, parents need to underline their abilities and hard work, expand their daughter's aspirations, and show them the endless possibilities that are available. It is equally important to have parents talk about injustice, not accepting the status quo, and standing up for women and explaining to their children that they can be part of historic change to build a better society. In school, teachers should build inclusive classrooms, introduce inspiring role models, and have students work on real-world problems that make science relevant and fun.

When I meet with middle-school and high-school-age girls and young women, I tell them that there is a place for them in science, that there is a real opportunity to bring new perspectives and ideas, and to create, innovate, and inspire new voices, communities, and the world—just as Greta Thunberg has done, and that science will be better with them.

All children should be encouraged to discover, explore, and try, and if they fail, that's okay. But they should learn from that failure. I explain to girls and young women that "any athlete, inventor, or scientist will tell you that many of the best lessons come from a bad day, a missed goal, or an outright failure and that there's no such thing as perfection. Be good to yourself." I repeat, "Believe in yourself, stay positive, and if you don't get it the first, second, or third time, try again, try harder, try smarter, and pick off one challenge at a time." These are important lessons to teach girls in science, math, and coding and to teach all students, and it should be part of the socialization process.

When aspiring young women are tired of not getting it right, I remind them of Marie Curie, Lise Meitner, Elsie MacGill, and a long list of others. I ask the students, "Did they give up, or did they dig deep and make a better future for us all?" They teach us that we are not alone and that scientists are not born, but rather develop through tough challenges. And I always finish with the same message: "The reason I tell you about these women is because what they accomplished is remarkable, and they did it at a time when women were hidden, the decks were stacked against them, and they had few options available to them. Yet they prevailed, they triumphed, and so can you, and through science, you can accomplish things that we cannot even imagine." Because they will need courage against hostile social norms to keep going, to persevere, to not be deterred, and to ultimately bring more young girls and women into science.

CHAPTER 4

Attracting Women to Science

F OR AT LEAST HALF a century, efforts to improve the representation of women in science have focused on educational reforms and individual programs to increase the number of women entering and staying in undergraduate and postgraduate education and training. Yet, just over a decade ago in 2011, Bayer Corporation polled over 400 STEM department chairs at the United States' top 200 research universities and those that produced "high numbers of female, African-American, Hispanic, and American Indian STEM graduates." While most STEM department chairs acknowledged that challenges existed for women—namely, stereotypes and a lack of role models—over 25 percent reported that there were no significant barriers for women or that "they did not know of any barriers." In addition, chairs did not recognize stereotyping as a problem for "African-American, Hispanic, and American Indian students."[237]

Therein lies the issue. A large percentage of chairs did not recognize that women and minorities faced challenges despite underrepresentation in STEM being an issue for decades, and despite 40 percent of the country's female and underrepresented minority chemists and chemical engineers reporting that they were discouraged from pursuing their STEM careers at some point in their lives.[238]

If something is not seen as a problem, there is no perceived need to act. Perhaps that is why only one-third of the chairs reported formal STEM diversity initiatives (although 66 percent reported that informal programs existed at their institutions) and over 30 percent of chairs did not know whether underrepresentation needed to be addressed by the highest institutional leadership. Academics and industry gave recruitment programs good and adequate grades, respectively. University chairs gave STEM programs to recruit and retain women a B- and a C+ for recruitment and a B- for retention of underrepresented minorities, respectively. Emerging STEM company CEOs and Fortune 1000 STEM company CEOs gave a C+ for both recruitment and retention of women and underrepresented minorities.[239]

On balance, recruitment and retention programs over the last decades have led only to incremental change, and today the proportion of women in computer science, engineering, and physics remains unacceptably low. The first real "leak" in women from STEM is at the juncture between secondary school and postsecondary education. Globally, more women are enrolled in universities than men: 112 women enroll for every 100 men, and graduation rates for women are higher.[240] Yet even with girls performing equal to or better than boys in STEM fields, fewer women study STEM at the postsecondary level and pursue STEM careers. Globally, only 18 percent of women pursue STEM compared to 35 percent of men.[241] "Only 7 percent of women choose engineering, manufacturing, and construction, compared to 22 percent of men." In the information, communication, and technology (ICT) fields, 28 percent of students are women and 72 percent are men.[242]

In 2018, Stoet and Geary published a paper about the "gender equality paradox" in STEM. They found that girls performed "similarly to, or better than, boys in science in two out of three countries," but that women were more likely to be under-represented in the sciences in more gender equal countries. The research not only grabbed headlines, but also drew criticism for its methodology and data, and the suggestion that biological differences – not social inequalities – drive the gender gap in STEM.[243] In 2023, Rippon pointed out that the Stoet study[244] referred to "endogenous interests" in determining career choice, "suggesting that a choice between science and humanities is somehow internally determined." Women have been dispelling claims that they lacked the talent and the interest in science for decades, and that work continues.[245]

In Canada, women make up 22 percent of engineering undergraduates, but account for just 13 percent of licensed engineers.[246] Women earned 36 percent of STEM bachelor's degrees, compared to 64 percent of males. Women's representation differed considerably among STEM fields: biological sciences (60 percent), general and integrated sciences (58 percent), "mathematics and related studies" (43 percent), physical and chemical sciences (32 percent), engineering (19 percent), and computer and information sciences (16 percent).[247]

In the United States, women earned 57 percent of bachelor's degrees but only 39 percent of STEM degrees in 2015-16.[248] Specifically, women earned the following percentages of bachelor's degrees: biological sciences (60 percent), math and statistics (42 percent), physical sciences (39 percent), engineering (21 percent), and computer sciences (20 percent).[249] Women earned 19 percent of computer degrees and hold 25 percent of computer jobs.[250] Women earned only one in five engineering degrees[251] and only one in five physics bachelor's degrees and physics doctorates in 2017.[252]

In the United Kingdom, women account for about one-third of students enrolled in mathematical and physical science courses (e.g., astronomy, chemistry, physics). But women account for only about a fifth of students in computing, engineering, and technology.[253]

This chapter addresses the challenges that undergraduate and postgraduate women face today, including the application processes that were designed to exclude, the "chilly climates" in the institutes and the lack of mentors and role models. In the face of these obstacles, I argue for systemic culture change to create belonging on campuses and in classrooms and to develop supportive mentors who build pathways to research careers.

* * *

The oldest university in the Western world, the University of Bologna in Italy, was established in the eleventh century, the oldest university in the English-speaking world Oxford in the twelfth century, and Cambridge in the thirteenth century, largely to produce male practitioners of the learned professions of divinity, law, and medicine and bureaucrats to run the church and state. As more medieval manuscripts are becoming available, we are learning more about learned women, who were not just anomalies, victims, but also community leaders.[254] Women were not admitted to the University of Bologna until the eighteenth century and although women had been studying at Oxford University for decades, it was not until 1920 that they could receive degrees for the first time.

The world's first university was founded by a woman. Fatima al-Fihri (b. 800 AD) founded a mosque that later developed into the famous al-Qarawiyyin University in the Moroccan city of Fez. "While ancient India's Taxila and Nalanda universities may date back further, al-Qarawiyyin holds the world record as it has continually offered education since its founding."[255] During al-Fihri's lifetime, she was known as the "mother of boys," likely because she took students "under her wing." In the 1940s, women were admitted to al-Qarawiyyin University, and in 2017, there was a push to encourage more training and professional opportunities for women through a prize created in al-Fihri's name.[256] Through the Middle Ages, education was not easy to obtain for anyone. This was especially true for women, who were seen as intellectually inferior. Monasteries offered education for privileged young men from the nobility and upper middle class, and convents provided opportunities for wealthy women who required some literacy to prepare them to be respectable wives and mothers. Some convent offerings included arithmetic, astronomy, geometry, grammar, reading and writing Latin, morals, music, rhetoric, spinning, and weaving.[257]

Annie Rogers was among the first women to graduate with full Oxford BA and MA degrees on October 14, 1920, more than forty years after taking her exams.[258] At Cambridge University, women were finally admitted in 1869 as Girton College became the United Kingdom's first residential educational institution for women. However, female students had to ask permission to attend lectures, had to be chaperoned, were not allowed to sit exams without special permission, and were not awarded degrees until 1948.[259] It was not until 1945 that the first Black woman, Gloria Carpenter, enrolled at Cambridge. She later helped "establish the

University of West Indies Law department, opening opportunities for Jamaican women to study law and to follow in her footsteps."[260]

From the moment women were "let in" to Cambridge, they were marginalized. They faced deep opposition, oppression, and violence—burning effigies of female students, fireworks, riots, and damage to one of the women's colleges—and their experience differed greatly from those of their male colleagues, who simply had to attend class.[261] Eventually, in the nineteenth and twentieth centuries, all-male faculties and departments began to make room, often begrudgingly, for women. At the same time, this inclusion slowly allowed for the enrollment of veterans, minorities, Indigenous peoples, and persons with disabilities. In 1918, the university Senate at Queen's University in Kingston, Ontario, voted to ban Black students from being admitted to the medical program. This racist restriction remained in practice at Queen's until 1965 and the policy stayed on the books until 2018, although it was not enforced.[262] The university formally apologized in 2019.

In 1827 in Upper Canada, a royal charter created King's College, which in 1849 became the University of Toronto. In 1873, the University's School of Practical Science was established, and in 1884 women were first admitted to the university, fifty-seven years after it was founded.[263] It took another forty-three years for the first woman to graduate from the electrical engineering program, and in total, two hundred years to reach one-fifth of engineering graduates and just over one-tenth of licensed engineers in Canada.[264]

At the end of the 1960s, Ivy League college presidents decided to admit women: Princeton and Yale in 1969, Columbia not until 1983. Their motivation was not to ensure equal rights and do the right thing, but to attract "the best boys," whose enrollment was declining in their schools because these students wanted to attend school with girls. When these conservative, prestigious colleges and universities announced their move to coeducation, they were met with "fierce resistance." Alumni pushed back hard with comments like "keep the damned women out."[265] It was argued that it made no sense to give valuable spaces at the schools to women because they were incapable of full-time work—their role was to give birth and raise their husbands' children.

In 1957, then-Senator John F. Kennedy agreed to participate in a debate at Hart House at the University of Toronto.[266] Female students were excluded; the building was restricted to men at the time. Women protested outside, and some brave female students disguised themselves in trousers and hats, entered the hall, and took their place near the front row. Unfortunately, a security guard noticed one of the women was wearing nail polish and they were all escorted from the building.[267] The future U.S. President said, "I personally rather approve of keeping women out of these places. It's a pleasure to be in a country where women cannot mix in everywhere."[268] It was not until the 1960s that more than a handful of women became tenured faculty members. Hart House became fully coeducational in 1972.[269]

Women still live in this heritage of exclusion. Men who have held power are not always keen to relinquish it and can find a myriad of excuses not to do so. As a result, conversations about equity and diversity have continued to be challenging in STEM departments, on campuses, and in the workplace. In my time, a constant refrain was that "women have not been in science or research that long and there just isn't the talent base." This is simply wrong: women have been in science, despite being deliberately excluded, unpaid, and held back, and the existing talent base is full of women who need to be included and promoted.

While girls and women perform equally well or better than men and boys in school, they fare worse in the workplace.

* * *

Let's explore the challenges that undergraduate and graduate women face today, starting with the male-developed letter of recommendation. While rigorous vetting of undergraduate and graduate students is an integral part of the academic system today, the practice of selective college admissions and exclusion in the United States is less than a century old. In the early 1900s, the Ivy League schools were not that expensive, they did not have ten to twenty times as many applicants as spots, and applicants were not thoroughly screened.[270] Shortly after Harvard College adopted the College Entrance Examination Board tests in 1905 as the principal basis for admission, "a crisis" emerged threatening the foundation of white, Anglo-Saxon, protestant, male privilege: 7 percent of the freshman class was Jewish, 9 percent was Catholic, and 45 percent was from public schools rather than prestigious, private boarding schools. Put simply, "academic prowess was leading to the wrong kind of student."[271] As a result, Harvard, Yale, and Princeton changed the definition of "merit" and put in place policies to ensure advantage to wealthy, white, protestant students, with an application system that prioritized blatantly subjective qualities—athletic ability, birthplace, family background, personality—over test scores.[272]

Multiple-page applications asked questions about religion, father's occupation, mother's maiden name, where a student's parents were born, and what if any changes had been made to a student's name.[273] A photograph became mandatory in the effort to keep out Jews, who by 1922 accounted for one-fifth of Harvard's freshman class. Harvard President A. Lawrence Lowell also required admissions officers to find out information about the "character" of candidates from "persons who know the applicants well." The letter of reference became compulsory and institutionalized.[274] Letters of recommendation, especially those from trusted sources such as alumni, headmasters, or teachers from leading feeder schools, were used to show that someone known to the college vouched for a candidate. It was a system built on privilege—an old boys' network that was meant to keep out women, minorities, and whoever else institutions did not want to admit.[275] When the letter alone was insufficient to exclude, interviews were added. In the 1950s, Yale was particularly seized with the character of "manliness" and even had a "physical characteristics checklist" through to 1965. The school measured each of its freshmen and boasted of the proportion over six feet.[276]

Today, letters of reference remain virtually universal for undergraduate and graduate school admissions and internships, fellowships, and grant applications throughout a person's educational and work life to determine who is chosen and who is not. Supporters argue that letters of recommendation offer valuable information in addition to grades, awards, and numbers of publications, and help assessors better understand special circumstances. Opponents argue that they do not help decision-making and contribute to bias and inequities; *implicit bias* refers to prejudices that develop unconsciously from beliefs and stereotypes and affect our daily actions, decisions, and how we perceive others.[277]

Today, there is a growing body of research that shows letters of recommendation feed networks of prestige and show bias against women; we appear to have come full circle, from the initial intent.[278]

In 2012, social psychologist Corinne Moss-Racusin and her team at Yale provided application materials for a laboratory management position to over one hundred science faculty members from biology, chemistry, and physics departments at research-intensive universities. The application details were identical except that half the faculty were given applications with a male name and the other half, applications with a female name. Female and male faculty rated the male applicant as significantly "more competent and hireable" than the identical female applicant. They also selected a "higher starting salary and offered more career mentoring for the male applicant."[279]

In 2016, Kuheli Dutt, assistant director for academic affairs and diversity at Columbia University's Lamont-Doherty Earth Observatory, and her coauthors showed that letters of recommendation may disadvantage women from the very start of their careers and that professors writing the letters may not even realize it. The authors analyzed the tone of over 1,200 recommendation letters submitted by individuals from over 500 institutions from over 50 countries to a highly competitive postdoctoral fellowship in geosciences at a top-tier U.S. university. They published their work in *Nature Geoscience*. Unsurprisingly, women were more likely to be described as "nurturing," "diligent," or a "team player," whereas men were more likely to be described as "confident" and "dynamic." If a department is looking for leaders, the choice of a referee's adjectives can make the difference in how a candidate is perceived and whether they get the position. The team found that women were only about half as likely as men to receive letters written in ways that portrayed them as excellent rather than merely good candidates. "It didn't matter if the letter was written by a woman or a man, or from Africa, America, Asia, or Europe. The results were universal."[280]

The Dutt study is important because it shows that universities continue to judge women and men differently in a STEM field and highlights a very specific process that can help reduce the number of women in science. While letters of reference are still in use, we should ask what training is being delivered at universities to address conscious and unconscious bias, what strategies are being developed, and how results will be measured. Importantly, the Dutt study shows that letters of recommendation are just one way in which gender biases emerge. After selection, women will go on to face many more hurdles, contradicting those who argue that bias does not happen in their department, field, or institution.

While I was a good student and graduated with distinction, getting letters for graduate school proved difficult. One professor commented, "Are you sure you want to go into research? You don't really look like a researcher to me. I don't think you would do well out in the field." It didn't matter what my grades were, in *his* esteemed view I did not look the part. I wore dresses and had long hair. But I was just as smart as any of my male colleagues and an athlete, as well. I threw double somersaults in the air from the floor and would go on to participate every weekend in running races and triathlons—including numerous Boston Marathons and half Ironmans. My strong hunch is that I could have beaten the potential referee hands down on the weight bench and on the track. But knowing that my professor had the upper hand and could determine my future, I calmly explained my reasons for wanting to do research and listened to him insist, "girls don't go to graduate school."

I would not let the professor stand in my way. I did find professors who were prepared to write to many schools on my behalf and I am forever grateful to them. One of my two women professors was incredibly supportive but also warned me that academia would be a hard life. She cautioned in the nicest possible way that, as a woman, it would be harder to get an academic job, harder to get funding, harder to get published, and that there could be challenges I couldn't even begin to imagine. She was right! I could not possibly have anticipated the sexism that would come my way. I was young, prepared, hopeful, and ready. My parents had expected the same of my brother and me; they had told me I could be anything I wanted. I wasn't supposed to face what I thought was my grandmother's and mother's fight, not mine. Today, I know we stand on the shoulders of giants—our great-grandmothers, grandmothers, mothers—and we owe them a debt of gratitude for each wave of feminism and hard-won rights they gained. But the fight is unfinished.

Once a young woman has been admitted as an undergraduate or graduate student, new challenges stand in her way. Recent research shows that, for STEM graduates who are women and/ or from underrepresented ethnic and racial groups, a lack of belonging can have a real impact on a critical career requirement: their publication record. Institutional structures and cultures are partially responsible. Aaron Fisher, a psychologist at the University of California, Berkeley, surveyed almost 350 graduate students in computer science, engineering, mathematics, and physical sciences at four universities in California. He found that "female and minority graduate students published the same number of papers as white male peers when they felt accepted by their mentors and peers, had clear departmental expectations, and felt prepared for their graduate classes." Some students were less productive because of their experience of being an outsider or being a minority in an environment that was designed for the majority.[281]

As we have seen, the drive to be social is central to human behavior, and a sense of belonging matters. Everyone wants to belong. There is no greater need than to feel part of something, to feel included, and when we are ignored or left out, disappointments can cause powerful reactions in the brain.[282] It can cause scars and some can last a lifetime. On my very first day of university, there was a warning from my male professors: "Look to your right, look to your left—they won't be here when you graduate." I had worked so hard during my last year of high school and I really felt that I had accomplished something and was now embarking on a wonderful new adventure. I was excited. I couldn't wait for my first classes and to hear from people who had spent their lives doing interesting research: learning from Indigenous scholars, unearthing ancient archaeological sites, and learning languages that were disappearing from the face of the planet. Those unwelcoming messages from these professors made it clear that everyone in the room had been at the top of their classes to get in the door and that, if we wanted to get anywhere, we would have to work hard. I never felt smaller or less significant than when sitting in the front row of that enormous lecture theatre. I couldn't wait to escape.

As I walked from the lecture theatre to the gym where I would try out for the university gymnastics team, I thought about what my former gymnastics coach would have said to welcome us to campus, even though he never had the opportunity to experience higher education himself: "Young women, I am so proud of what you have accomplished to be here; you are all champions in your own right, you belong here, and we're going to do great things together. Now, don't get me wrong, there will be hard work ahead, but I need you to dig deep, and I'll be here to support you every step of the way to achieve your goals." And with those words ringing in my ears, I walked into the gym, got ready to tumble, and made the team.

Students want to know that their professors care. They are no longer willing to put up with institutions, departments, and professors who do not understand the principles and arguments for inclusion and belonging. Thankfully, in the times that followed, my university life changed for the better. An inspired geography professor—Professor Kenneth Hare, award-winning scientist, university dean, president, and chancellor—provided a spark for a new path of research. He took us on a two-hour journey of our planet from the frozen poles to the lush tropics and the grasslands of Africa. He shared his decades of fieldwork, opened our eyes to the big questions of science, and shared how our planet might change in our lifetime. At the end of our class, he challenged us to pay attention to the world around us: look in every tree and notice who is nesting or hiding in it. Turn over every rain-soaked leaf and see who is lurking on the underside. Really pay attention and learn to see the beautiful world we live in. I knew on that very day that I wanted to do research and to teach at the university. That had become my dream.

One professor in two hours had managed to accomplish what many of his university peers had failed to do—make me curious about science. I had not been part of any STEM pipeline. My life until that moment in university had been art, dance, music, and gymnastics with a dream of becoming an artist and a dance teacher. Incidentally, I still love to teach dance and coach sport. Growing up, I knew no female scientists, had no high school volunteer science opportunities or science mentors, and felt no proficiency in science. No one ever questioned why I dropped two science and two math courses or encouraged an interest in STEM fields. Professor Hare asked us to explore and discover the world around us every day and this had an impact on me. It changed the way I interacted with the environment, making me see things I wouldn't have noticed before.

The experience also made me realize that there was a real difference between who led the university, who served as administrative support, and who was studying. Men ran departments, men were professors, and men were teaching assistants (TAs), while it was largely women who served in administrative roles. Yet in the first year, class enrollments were mostly split equally between women and men. I was often made to feel like an intrusion if I dared to knock on a professor's door, and I quickly found that it was the women administrative staff who provided support when I had questions about courses, exams, or the future.

One professor regularly threw a fake brick at students to wake them up and intimidate them. I was surprised that the professors had no better techniques to maintain discipline in the classroom. When visiting one of my TAs in his office for help, I was put off by the graphic penthouse pin-up staring back at me. He was twenty-four years old, straight out of undergrad, but someone who was overseeing students and planning to do a PhD. He would have even more responsibilities to young people and more power over their grades and career paths in the future. Feeling I didn't belong, I opted out of classes that required labs or tutorials. My university experience was diminished from that day forward.

Perhaps more damning, a psychology professor showed slides of dead, hanging, homosexual men, each with an electrical wire strung from one nipple to the other "in an attempt to heighten their orgasm." What dignity was afforded to these men in death and how did those pictures affect 2SLGBTQI+ students in the class—indeed, all the students in the class?

It makes me wonder what might have happened if the university community had been better when it came to women, researchers from racialized backgrounds, Indigenous scholars, scientists and researchers with disabilities, or those from the 2SLGBTQI+ community? Would it have taken to the 1970s for breast cancer to have become an issue of national concern? When HIV/AIDS began, would the 2SLGBTQI+ community have been painted as alien, something to be feared? Would the Reagan administration still have been slow to act?

Our sense of belonging can be undermined by structural discrimination, bias, racism, and unfairness. These factors are real. It was encouraging when the journal *Nature* came forward in 2022 and acknowledged how it contributed to science's discriminatory legacy. In 1904 and 1908, for example, the journal published and printed speeches by statistician Francis Galton on eugenics and how communities could start their own local associations favoring "fit" families for citizenship. Eugenics was racist and wrong. But it was backed by some prominent scientists, politicians, and lawmakers and offered those in positions of power a theory and language to support their biases against those they feared most. It became an international movement. In the United States, eugenics led to the forced sterilization of so-called "undesirable" populations: immigrants, minorities, poor people, unmarried mothers, the disabled, and the mentally ill.[283] In Canada, a 2022 Senate report showed that women and girls were manipulated, threatened, or sterilized without their knowledge in an effort to subjugate and eliminate First Nations, Métis, and Inuit, as well as Black and racialized women, and persons with disabilities.

Each scientific article from history that contributed to sexism, racism, colonization, and inequality must be identified so that we can gain a full picture of how research contributed to the exclusion and subjugation of people and how these historical undercurrents are perpetuated today. Each university, department, and institute should also look at its history, who was excluded and why, and who were the champions who fought for change and were they ever recognized? Departments and institutions that are working on these histories can ensure that students aren't made to feel that they are imagining the hurts, slights, and wrongs, or that they are alone without help. While these longstanding issues are addressed, it is important to have peer groups with safe spaces where all women can talk about what they are experiencing and what collective action might be required.

It also helps to have supportive professors and supervisors who can help chart a path forward, providing structure, clear expectations, encouragement, accountability, and guidance. Unfortunately, not everyone has a great supervisor.

I was lucky. I had the best. Professor David Sugden, an award-winning scientist who visited Antarctica over fifteen times and made seminal contributions to the glacial and climate history of the southernmost continent, met with me every two weeks during my PhD. We went over the work from the previous meeting and what was expected at the following appointment. He asked about what was going well, what was challenging, and how he could better support me, and I left feeling like I mattered and that I had a place not only in our department but also in our university, despite being in a country far from home. Some friends at other universities referred to their intimidators, tormentors, tyrants, and, in some cases, harassers or worse. All of us at the university have heard these horror stories of absent, abusive, and negligent supervisors who belittle, ignore, obstruct, sabotage, steal credit, and sometimes have inappropriate relationships. A handful of times during my life at the university, students reached out to me for help following the devastation of one of those relationships with a clear power differential. I got them help that very day.

None of us should be surprised by these negative interactions, despite all the work that has been done to improve the graduate experience. Scientists advance in their careers based on their research, not their teaching, supervising, mentoring, or communicating. University tradition is that the possession of a doctoral degree alone qualifies a scientist to teach the next generation of scientists. We should ask if any of the training for new PhDs include aspects such as legal responsibility, ethical responsibility, and ideas about leadership, teamwork, and citizenship. Does training include equity and diversity, hiring practices, people management, and the management of research funds provided by taxpayers? Was information provided on how to teach, how to maintain discipline in the class, or how to mark and do justice to students' hard work? On big campuses, there is often a business school, an ethics department, a law faculty, and a teaching faculty from whose expertise the whole institution could benefit. And if new PhD graduates hadn't received any of this training, why should we expect her, him, or them to have these skills that can improve lectures, classroom, and research culture, and help retain women?

Elementary and secondary school teachers go through intensive teaching theory, practice, and training, and sport coaches go through theoretical and practical training and exams to work with students. Yet when it comes to postsecondary education, no training is required. Universities provide institutional apprenticeships that are meant to train the next generation of researchers and meet the schools' labor needs. Because they learn through apprenticeships, graduate students often teach and treat the next generation in the same way that they themselves learned. As a result, the culture of a particular discipline or even a department or faculty is perpetuated. If it is acceptable to disparage women, utter comments like "gays don't belong in the hard sciences," share inappropriate cartoons and jokes, undermine colleagues, work terrible hours, or insult or scream at or shame students to "toughen" them up for the "brutal realities of science," these behaviors are likely to continue.[284] No professor or teaching assistant should ever demean or yell at a student. It is abuse, and it must stop.

For the first century of medical residency, 120-hour work weeks were normal.[285] Residents were required to spend long hours with patients—up to forty-eight at a stretch—observing and treating the person and understanding the progression of the disease to become good doctors. There were mistakes, patients died, and good doctors left the field.[286] The treatment was abusive and outdated. It needed to be replaced.[287] A lot of similarly abusive practices need to be replaced.

I would like to see every new doctoral student who is hired to teach young people complete mandatory ethics, leadership, management, and some teacher training when they join a university. I would also like to see faculty have mandatory training each year to remain current with new laws, regulations, and codes. Professors should be trained to work with young scholars who do not look like them and who do not have the same lived experience. Far too often, diversity and inclusion are reduced to the numbers. Those numbers should be the starting point, not an end in themselves. We need to make sure that everyone has a voice, an opportunity to be heard, and that there is appreciation for everyone's background and the perspectives they bring. This will help women be their authentic selves in class, the lab, or the field and not have to apologize for having different ideas, different interests outside of science, or for having a family. Creating a sense of belonging and choosing to include is the right thing to do; we must let the next generation know that the sky is the limit for all of them, not just some of them.

The term "chilly climate" was coined by Roberta Hall and Bernice Sandler (1982) to describe the countless small inequities in the classroom that, taken together, create a forbidding environment.[288] These behaviors and comments can be unconscious or unintended. They are often referred to as microaggressions, such as correcting, dominating, and interrupting. But they can also lead to blunter, starker incidents, such as having a professor explain that "the only reason a woman made it to the top was because of who she was sleeping with."

Recent, ignorant comments by Nobel laureates James Watson and Tim Hunt highlight the challenges that women, and particularly, diverse women in STEM fields face. In 2007, Watson, a member of the 1962 Nobel-Prize-winning team of Francis Crick, James Watson, and Maurice Wilkins who discovered the molecular structure of DNA, made ugly, racist comments about the continent of Africa and Black people.[289] Disturbingly, the comments were not Watson's first example of gross misconduct. The Watson and Crick model of DNA was based in part on the x-ray crystallography work of Dr. Rosalind Franklin whose contribution was downplayed and denied for years. Franklin suffered gender discrimination, anti-Semitism, cronyism, stolen data, and a cover-up of the theft. She died of cancer at age thirty-eight, five years after Watson and Crick's discovery. She was not named a Nobel Prize winner, as the Prize is never given posthumously. Moreover, Franklin's contribution was not even acknowledged by the men during their Nobel speeches. It was a shameful snub of an outstanding woman scientist and her work.[290]

In 2015, British Nobel laureate Tim Hunt "came under fire" for saying women shouldn't work in laboratories because their presence leads to romantic encounters and therefore harms science. Specifically, Hunt opined on the "trouble with girls" in science labs: "You fall in love with them, they fall in love with you, and when you criticize them, they cry." Women scientists rightfully hit back hard on Twitter with comments using the hashtag "distractingly sexy," while showing their work in cell cultures, coding, environmental engineering, and other disciplines.[291]

If these Nobel laureates' outrageous, ridiculous, and shameful comments are what some of the biggest names in science believe and think it is acceptable to say publicly, what is said behind the scenes, who is chosen to work in labs, and what treatment must they endure? If we do not face chilly climate, bias, discrimination, and stereotypes head-on, we will continue to lose women in postsecondary education. We will have to continue to tell them what we have told them for a century: keep persisting, ignore the bad behavior, it will get better, your time will come, and isn't it wonderful just how much progress women have made.

Often, academics who belong to the privileged group have limited awareness of chilly climates, including even woke colleagues who see themselves as allies. In 2017, Jennifer Lee, then a sociology major at at Dartmouth College, and Janice McCabe, an associate professor of sociology at the same college, undertook a study to find out if a chilly climate still existed in the classroom. They wanted to know if the behavior had changed since the work of American University professors Myra Sadker and David Sadker (Hall et al., 1982, 1995), who together gained national reputations for confronting gender bias.[292] Lee observed ninety-five hours of classes in nine courses in the humanities, natural sciences, and social sciences, with five of the classes being taught by women and five of the classes having mostly female students. The professors' and students' body language, exact words and context, and frequency of engagement were coded to allow pattern identification. They found that the men spoke 1.6 times more often than the women, spoke out without raising their hands more than women, were more likely to use more assertive language and interrupt others regardless of the gender of the professor and makeup of the class, and were more likely to have longer conversations. By comparison, women were more hesitant to speak and more likely to use "apologetic language." The research suggested that gender hierarchies persist and women's voices may still go unheard, despite being the majority in U.S. colleges.[293]

We know that from an early age, gender norms create social roles and expectations of how women and men should behave[294] and that by college there is already evidence of manologues, mansplaining, and manterrupting, which later play out in the workforce. In fact, interruptions even happened at the U.S. Supreme Court. Professor Tonja Jacobi, Emory University Law School, and Dylan Schweers, a student at Northwestern Pritzker School of Law, showed in 2017 that male justices interrupted female justices approximately three times as often as they interrupted each other during oral arguments. Over time, women reduced their chances of being interrupted by adopting male speech patterns: for example, forgoing polite pleasantries such as saying, "excuse me."[295]

The interruptions of female justices have led to a positive change in the form of new Supreme Court rules.[296] We need to address this earlier, however. Ideally, throughout school classrooms, and preferably before college, so that it does not carry into the next generation. When Dartmouth "professors took proactive measures to engage all students, such as outlining the expectations for participation and calling on female students who had not had a chance to speak, this created a more equitable discussion."[297] Both professor–student interactions and student–student interactions contribute to classroom climate. A positive classroom climate is associated with the creation of a safe and welcoming environment, instructor kindness, encouragement of inclusion, mutual respect, and participation, shared interest in course material, peer acceptance, and peer support, and intervention when students exclude or marginalize others. More specifically, do all students ask questions? What effort is made to draw out quiet students? Does the instructor take questions alternately among women, men, and gender-diverse students? Is feedback provided equally, and how can answers be strengthened? And does the professor intervene when students belittle, disrespect, or interrupt other students?

As a minister in government, what I heard loud and clear as I crisscrossed the country meeting with undergraduate and graduate students was that the climate in STEM remains chilly. Young women complained of young men "sucking all the oxygen out of the room," not being able to get supervisors for thesis courses, sometimes because a professor was uncomfortable having a woman in his lab, not being able to get letters of recommendation, and "having to be twice as good as the men just to get noticed."

When I was an undergraduate, a part-time student dared to complain that a senior, accomplished, male-tenured professor of engineering was staring at her while she swam at a university pool. He made her feel uncomfortable. The professor later admitted to looking at women while they were swimming. In his own words, he said, "I observe women, I enjoy women, I photograph women but I do not consider that I leer." An accomplished photographer, he also showed pictures that he was asked to take of the synchronized swimming team. The university's newly formed sexual harassment board found that the professor's actions constituted sexual harassment and some board members thought his pictures were "salacious and pornographic." He was found guilty of "prolonged and intense staring of a lewd or lascivious nature" and banned from the campus pool for five years.[298] At the time, that decision was controversial. Some condemned the fact that looks could be "illegal," and that there was no physical or verbal contact between the professor and the student. Moreover, many saw this behavior as part of the normal interaction between men and women and an infringement on personal liberties. Among my female fellow classmates, the question was: "if he's staring at women in the pool, what's he doing in his class, and why is there no line there?"

At nearby Queen's University in Kingston, Ontario, women were subjected to break-ins and vandalism, humiliating remarks, like "show us your tits," and men simulating sex by performing push-ups over them on the football field. In 1989, male students at Gordon House residence put up signs that mocked the "no means no" campaign being run by several campus groups. The men's signs read: "no means maybe," "no means have another beer," "no means tie me up," "no means kick her in the teeth," and "Down on your knees bitch."[299] In response, principal David Smith created the Principal's Working Group on Gender Issues to change Queen's sexist culture. The group advised cracking down on sexist clothing with slogans such as "Lick it, slam it, suck it," rugby songs with pro-rape lyrics, and the student newspaper *Golden Words*, published by Queen's Engineering Society since 1967, which included a "Slut of the Week" feature.[300]

These stories show where we have come from. Yet these details are usually whitewashed by self-congratulatory promotions such as

"100 years of women." Institutions tend to celebrate their long and illustrious records and important milestones with respect to women while ignoring the exclusions, protests, and efforts by brave women to make change.

According to Åsa Regnér, former deputy executive director of UN Women, "the number of girls interested in STEM almost doubles when they have role models they can name and identify with."[301] Fortunately, more women scientists are teaching and researching than ever before. They can help encourage, engage, and inspire young women to achieve and help counter the social norms, expectations, and gender bias they face in their education.

I know what it feels like to be both a student in a department with few women and one of only two women faculty members in a department. I had only two women professors in four years of study. And there was no one in my department with whom to share anything as nebulous as hopes and dreams. I talked instead to my mother, a high school gym teacher, and my father, a caretaker and talented baseball pitcher, about a career in research. In our family, the expectation was that you excel in school and excel in sport and, believe me, sport mattered more.

In the male-dominated fields I studied as an undergraduate, women's studies were often disparaged in our lecture halls. We heard about the "great Louis Leakey," a man who traveled frequently to the United States to lecture, raise money, and often left the impression that he, personally, had made his wife's discoveries.[302] We also heard that "Mrs. Leakey" didn't finish high school, loved her dogs, smoked cigars, was overly fond of whiskey, and was responsible for her husband losing his Cambridge post because they lived together while he waited for his divorce. In fact, the crowning triumphs of Mary Leakey's career included finding 400 skull fragments of an early hominid, which she carefully pieced together over a period of eighteen months, and finding 3.6-million-year-old hominid footprints at Laetoli in Tanzania, which proved early humans were bipedal.[303]

World-renowned ethnologist, conservationist, and activist Jane Goodall was dismissed by professors as not having a PhD when she began observing chimpanzees in Africa. We also heard that she was chosen only because of Louis Leakey's fascination with her blonde hair and long legs. She was ridiculed for naming chimpanzees and for giving them "personalities."

Marine biologist and nature writer Rachel Carson suffered a similar fate. We heard from some male professors that she did not have a PhD and "had written a book." The book was the ground-breaking *Silent Spring* that took on the chemical industry in the United States and detailed how the unregulated dumping, dusting, and spraying of the pesticide DDT jeopardized countless bird species.[304] In response to the book, Carson was portrayed as a hysterical woman, a "bunny hugger," a communist sympathizer, and "a spinster who liked cats." Regardless, her book ignited the environmental movement and prompted the U.S. Government to act with landmark environmental protection laws and the creation of the Environmental Protection Agency. Soon after the publication of her book, she testified before Congress while she battled breast cancer.[305]

Seeing women faculty members and hearing about powerhouse women as an undergraduate would have let me know what was possible. Every student should learn about Nobel laureate Marie Curie and her remarkable accomplishments: two Nobel Prizes, her trips to the World War I battlefront, and her creation of radiology vehicles or "petite curies" to save soldiers' lives. Students should know about her relentless resolve, her tireless pursuit of excellence and dedication to improve life for others, and her refusal to accept the status quo. Despite her accomplishments, she never achieved the status of Carl Linnaeus, father of taxonomy; Charles Darwin, father of evolution; and Gregor Mendel, father of genetics. Was it because she was the first woman in France to earn a PhD in physics, spent "too much time" in the lab rather than with her daughters, or because her work was often attributed to her husband?

At Curie's first Nobel ceremony, the president of the Swedish Academy, who administers the prize, quoted the Bible in his remarks about her research: "It is not good that man should be alone, I will make a helpmeet for him."[306] Were slights like these because Curie was an outsider, from the wrong social class, a woman scientist who dared put herself forward for the French Academy of Sciences? Was it that she had "tarnished the name of her good husband," or some combination of all these strikes?

Talking to undergraduates and graduates today, I find they cannot rhyme off the great women scientists like Curie and Lise Meitner—the codiscoverer of nuclear fission who was nominated forty-eight times for the Nobel Prize and denied each time. Meitnerium is named for her. Moreover, Meitner refused to have anything to do with the Manhattan Project to develop a functional atomic weapon during the Second World War. The press wrongly labeled her the "Jewish mother of the bomb."[307] More recently, she was among the many female scientists unfairly written out of history by the blockbuster movie *Oppenheimer*.[308]

What I hear from high school and university students is that they are still learning the same long list of male scientists I learned. We need a database of women and diverse scientists and their accomplishments so that every young person knows science is possible and that there are people like them in every field. The Lost Women of Science Initiative is working to that end, telling the stories of women who have made ground-breaking achievements in their fields. Dr. Jess Wade, a British physicist, is making it her mission to tell the stories of women scientists on Wikipedia. Since 2017, she has written more than 2,000 Wikipedia entries for female scientists and engineers and those from under-represented backgrounds and minorities whose accomplishments were not documented on the site.[309]

Databases and stories matter, but professors need to share the accomplishments of women scientists in the classroom.[310] Schinske et al. (2017) undertook to study whether student assignments about diverse scientists would allow young people to better see themselves in STEM. So-called "scientist spotlights" were short class assignments that presented diverse perspectives on who scientists are, how science is done, and scientists' personal stories. In each spotlight, students reviewed both the scientist's research through journal or popular science articles and the scientist's personal history through a blog, podcast, or TedTalk. They brought scientists to life. They presented science as an intensely human experience, a creative endeavor that involves coming up with and testing new ideas and new methodologies, and producing new results and new knowledge that might benefit everyone. Unsurprisingly, the researchers found that students who completed the scientist spotlights assignments shifted their thinking away from stereotypes of scientists and increased their interest in science. When Marie Curie is just a name, she is an impossible standard to emulate and reach. But when her full story is known, she is human just like the rest of us—living, loving, and striving—and we can identify with some part, or maybe many parts of her journey. She becomes an inspiring role model not just for women and girls, but for all.

More than role models, women need mentors who are going to be there to advocate, champion, and sponsor, particularly in departments with only a few women. In 1985, Jill Bowling and Dr. Brian Martin, both of the Australian National University, described academic science as a patriarchal system that incentivizes masculinity in its very structure. Although their work is almost four decades old, the underlying attitude that they described persists today.[311] The philosophy of exclusion is framed as "fit," as in whether or not the student will fit in the department, or maintaining excellence, as if women would dilute excellence, or maintaining the integrity of the discipline.[312] Research shows that active and effective mentoring can have a positive impact on equity, diversity, and inclusion in STEM; academic achievement; retention; degree attainment; and career success and satisfaction. Yet, students from underrepresented groups continue to receive less mentoring than well-represented peers.[313]

A 2019 report from the U.S. National Academies of Sciences, Engineering, and Medicine defined *mentorship* as "a professional, working alliance in which individuals work together over time to support the personal and professional growth, development, and success of the relational partners through the provision of career and psychosocial support." According to the report, there is a growing body of research about how to create, develop, and sustain inclusive, successful mentoring relationships, but there remains a gap between this evidence and current practice in universities. The report also showed that institutions have largely left mentoring to develop naturally and on an "ad-hoc basis." STEM mentoring in the United States does not, on average, follow best practices, and mentoring does not receive the "attention, evaluation, and recognition" given to research and teaching at the university.[314]

It is time that institutions take advantage of evidence-based, formal mentoring to improve recruitment, retention, and paths to research careers of all women and limit negative experiences. To ensure results, mentoring should be rewarded, and evaluative programs need to be put in place to ensure continuous improvement for our students.[315] We need mentors who are not only willing to advise, provide career and psychosocial support, and be there on the good days and hard days, but who are also willing to fight stereotypes, bias, and discrimination and refuse the role of gatekeeper and guardian of the status quo. If we get this right, we build strong relationships among mentors, mentees, and women across the career-span, disciplines, and institutions.

After decades of debate, handwringing, and millions of dollars invested, we have failed to solve the problems that keep women out of science.

While there is now a shared understanding that gender bias has affected research outcomes and women's health, there are researchers who still do not believe that women face challenges in science and tenure makes rooting out their attitudes difficult. In some quarters, these scientists have the backing, voice, and ability to determine who is admitted to a field and who is not. There are also organizations that believe that when women and diverse students and faculty are included, a new form of scientific excellence is required—an inclusive excellence.[316] Excellence is excellence, why the qualifier? We continue to treat women scientists like a special interest group rather than one representing half the population, and we continue to require women to fight for each incremental step forward. Women should not be made to feel they are peering in from the fringe, fighting to be seen, to be heard, and to be included.

Addressing underrepresentation in undergraduate and graduate classes requires action on all fronts by all departments and institutions. It requires looking at departmental and institutional data, determining where the shortfalls are, what processes need to be put in place or strengthened, who has oversight, and who is accountable for change. We know the problem, we also know what works, and yet there continues to be resistance in academia. Looking to the future, there is an ambitious goal to land a lunar mission by 2026. While there are questions as to who will be the first woman on the moon, equally important questions are who will make up half the mission team and mission control teams? It is the mandate of universities and colleges to prepare women for lunar missions, expeditions, and opportunities to run technology companies. That obligation begins with the creation of a sense of belonging, leading to good undergraduate and graduate STEM experiences, ensuring that women want to pursue careers in science.

CHAPTER 5

Challenges In and Out of Academia

F OR HUNDREDS OF YEARS, women were believed to be the inferior sex, "their bodies weaker, their minds lesser, and their roles subservient,"[317] so it is no wonder that once they were let into universities, they were denied degrees, equal pay, and professorships while being ridiculed as hysterical and hormonal and judged on their looks. In recent decades, the "leaky pipeline" metaphor has been used to describe how women have been abandoning STEM programs. Initial factors to explain the decline of women's involvement in STEM have proven false: they had the wrong biology, the work went against the natural order and was too difficult, women lacked the mettle, or they couldn't balance their home life with their work life. These notions were not only wrong but they also put the blame squarely on women rather than on the barriers to their success.[318] Women disproportionately chose to leave a career for many reasons, including insensitive interactions, isolation, and a lack of mentors and sponsors.[319]

Today, across many STEM fields and career stages, women

continue to report that they need to work harder to be successful and, in turn, experience a lower sense of belonging compared with their male colleagues.[320] Misogyny is by no means limited to university, but it remains a systemic obstacle that women face in the modern academic setting.

In the United States, the debate about whether women should have access to higher education started in the 1830s and lasted a century. Some claimed that university life would destroy a woman's femininity and her role as a wife, mother, and homemaker.[321] Others claimed that higher education for women would be acceptable provided it was carried out separately from men to prevent moral breakdown.[322] Two private colleges, Oberlin and Antioch, allowed coed education prior to the Civil War. But, in practice, only some classrooms were mixed, extracurricular activities were closely monitored, and an 1837 Oberlin policy excused female students on Mondays so that—wait for it—they could do the male students' laundry.[323] Because women had been excluded from higher education for centuries, their gradual and reluctant admittance was often met with protest, disdain, and ridicule.

In 1869, Mary Anderson, Emily Bovell, Matilda Chaplin, Helen Evans, Sophia Jex-Blake, Edith Pechey, and Isabel Thorne made history by becoming the first women to be admitted to a degree program at a British university.[324] The seven trailblazers chose to study medicine at the University of Edinburgh where they were forced to attend separate lectures from their male peers, denied access to the Royal Infirmary's wards, and required to pay more for a lesser experience. When these seven women dared to sit for their anatomy exam at Surgeons' Hall, they were met by an angry mob who caked them in mud and verbally abused them; the gates were shut in their faces. Understanding janitors and peers helped to provide them with safe access to the exam hall whereupon a live sheep was released into the room to disrupt the exam.[325] Ultimately, the women were rejected four years after they were admitted, and the University of Edinburgh denied them a graduation. However, the publicity around their plight "put the right of women to study on the national political agenda." The first women graduated from the University of Edinburgh in 1894 and the first doctors in 1896. In 2019, 150 years after they matriculated, the Edinburgh Seven were recognized posthumously by being awarded honorary degrees.[326]

The struggle was no different in colonial Canada. In 1836, the Upper Canada Academy in Cobourg, Ontario, began to offer secondary school classes that were coed. Five years later, the academy became a degree-granting postsecondary institution, and its name was changed to Victoria College. Egerton Ryerson took over as principal and excluded women, who thereby "lost their first chance at postsecondary education" and would not be admitted to Victoria College again for nearly forty years.[327] Ryerson did not stop there, gaining further notoriety as one of the primary architects of Canada's residential school system, where he helped to seal the fate of an estimated 150,000 First Nations, Métis, and Inuit children for more than 150 years.[328] No school should have had a cemetery next door or, rather, an unmarked burial place. And no child should ever have been awakened in the night to bury a brother, sister, or another child. These children were forcibly taken from their homes and sent to religious schools where they endured spiritual, cultural, emotional, physical, and sexual abuse, and thousands of them never came home to their families. Survivors had long testified that children who died at these schools were buried in unmarked plots.[329] These graves are now being rediscovered throughout Canada.

While Ryerson's shameful legacy left long-lasting and damaging effects on people and the country, there were other men along the way who helped to crack open the door for women. Such as the janitors at Surgeons' Hall and William Houston, who served on the University of Toronto's Senate (1882–1904). Houston not only helped women get the municipal vote in Toronto but also was a tireless advocate for women's higher education. Still, trailblazing women were forced to dress as men to gain admission to postsecondary education and endured insults, protests, and strikes. Some men lost no opportunity to make life as unpleasant for women as they possibly could.[330]

In 1880, women were allowed to enroll at the Royal College of Physicians and Surgeons in Kingston, Ontario, a precursor to Queen's University's medical school.[331] Physiology professor Kenneth Fenwick enjoyed sharing insulting anecdotes about women, to draw attention to the few he had in the class, prompting laughter from their male peers. In an 1882 lecture on the larynx, he compared the pitch of a woman's voice to that of an ape. When the women attending the class lodged a complaint, Fenwick and the male students countered that any change to his lecture would restrict academic freedom.[332] And there it was. One male professor refused to treat all students equally and the specter of academic freedom was invoked to shut down women, their voices, and science. While the Royal College's authorities first backed the women, they later caved, particularly when some of the men threatened to leave the college, and the women who initially enrolled were expelled in 1883. In the same year, they, along with supporters, formed the Women's Medical College, but a lack of funds and low enrollment led to the closure of the college in 1894. Women students at Queen's would have to wait almost another fifty years to take classes again at the faculty of medicine.[333] While the medical school at Queen's University had been the "first in Canada to admit women to its undergraduate courses," it was also the "first to exclude them" and "one of the last to readmit them."[334]

The Women's Suffrage Society sent a petition in 1882 to the Ontario provincial legislature asking that women be admitted to the University of Toronto. The president at the time chose "steady passive resistance" and voiced concerns about the need for women's lavatories, a women's superintendent, and a gym, which was built only after a forty-year struggle. At McGill University, women serious about their studies were ridiculed as "bluestocking" or unfeminine; they were also thought to be too involved in extracurricular, non-academic activities, and unserious. Beginning in 1903, women earned PhDs at the University of Toronto and over the next twenty-five years, twenty-eight of the thirty doctoral degrees awarded to women were in the sciences. Yet in 1926, men were still deliberating whether "woman has more than come into her own" at Hart House, which was restricted to men only, echoing an 1875 debate among men on whether it was "advisable that women receive a university or professional education."[335]

If advancement for women in science was slow, progress for female faculty was glacial. In 1931, after a request by the premier of Ontario, the president of the University of Toronto submitted to the board of governors a list of married women the institution employed and whether they could be replaced by "other persons." Two months later, the board resolved that it was undesirable to employ married women unless "such persons" required money to support their families. In the 1930s and 1940s, certain department heads worked to maintain all-male faculty and refused to grant women tenure.[336]

It was not until the 1960s that questions relating to permanent academic employment and tenure for women were even examined in Canada. Women hired as faculty received lower pay than men in virtually all departments and regularly found themselves to be the "only woman in the room." They did not receive proper research funding and did not have opportunities to advance.[337] In 1970, women made up 13 percent of total full-time faculty, and "the gap in median salary between men and women (in constant inflation-adjusted dollars) was widest in 1990 when men earned about 25 percent more than women."[338]

Meanwhile, the fight for women to be fully accepted on university campuses raged for decades in the United States, with their numbers steadily climbing through the 1920s. The Great Depression of the 1930s and the start of the Second World War then steered attention away from education. After the war ended, the country focused on the sacrifices that men had made in defending the country, with the result that women who had filled professors' roles in the interim were "overshadowed" by the men as they came home.[339]

In 1969, historian Margaret Rossiter was one of only a few women enrolled in a history of science program at Yale University. She dared to ask at a regular informal meeting of professors and students if there had ever been any women scientists. The answer was "No. Never. None." This was truly astonishing: in 1969 at Yale, a flat denial of the very existence of trailblazing women scientists. No mention of Lise Meitner, no Rosalind Franklin, or no Irène Joliot-Curie. When contrary evidence of the achievements of women scientists was presented to the group—namely, Marie Curie's work—she was deemed to be inadequate by the professors, despite having won the Nobel Prize twice. She had been a mere helper to her husband. Rossiter chose to make research on women in science her career. She worked for over a decade on her 1982 landmark book, *Women Scientists in America: Struggles and Strategies to 1940*. Her pathbreaking writing brought to life women astronomers, biologists, chemists, psychologists, and their triumphs, all of them hidden and suppressed by men and overlooked by history.[340]

The fight for admittance and recognition of women in science is part of a larger struggle. In August 1970, 50,000 women held the first women's strike for equality, blocking New York City's Fifth Avenue during rush hour. *Time* magazine reported that this movement emerged because "virtually all of the nation's systems, industry, unions, the professions, the military, the universities . . . [are] quintessentially masculine establishments."[341]

Women have been leading a discussion on "sexism in science" for well over fifty years. A quick Google search reveals over 140,000 results. Here is just a sampling of academic headlines over the decades: "Sex and Science" (1979); "Sexism in Science" (1981); "Sexism in Science" (1987); "Sexism and Scientific Research" (1997); and "Taking the Long View on Sexism in Science" (2015). Legions of women have gathered data and worked to prove that the issues they face in science are real and demonstrable. In the late 1990s, MIT molecular biologist Nancy Hopkins and other women at MIT took out a tape measure and showed that while senior male professors had labs averaging 3,000 square feet, senior female professors had on average just 2,000 square feet, the equivalent of junior male faculty members.[342]

The fight is ongoing. American women in STEM are more likely (50 percent) to report discrimination in the workplace than non-STEM workers, and much more likely than men to report workplace discrimination (41 percent and 19 percent, respectively). Moreover, women in male-dominated STEM jobs report higher rates of discrimination based on sex: 78 percent to 44 percent, respectively.[343]

Over 70 percent of women reported experiencing bias and microaggressions and almost one-third of women (29 percent) report being treated as if they were not competent, compared to only 4 percent of men.[344] Women often report that men tend to explain issues to them in their areas of expertise. (I cannot count the occasions, during my time as a minister of science in the Canadian government, when a senior male executive explained to me what a master's or PhD degree was, or "just how far women had come in science," or suggested that women's further promotion in science would damage research irreparably.)

Fifteen percent of women report having less support from senior leaders, compared to 7 percent of men.[345] Women also report insensitive interactions, isolation in the workplace, lack of mentoring and networks, lack of role models and sponsors, and sexual harassment.[346] I have heard, loud and clear, that women in science are tired of unfair work environments, having their accomplishments or ideas credited to someone else, having their judgment questioned, and having to work harder to demonstrate competence than their male peers.

For a range of reasons, women researchers tend to have "shorter, less well-paid careers." Their work remains underrepresented in prestigious journals and they are far too often passed over for promotion. Women tend to receive smaller research grants compared to their male colleagues, and while they represent one-third of all researchers, just over one-tenth of members of national science academies are women.[347] Only 6 percent of Nobel Prize laureates are women.[348] A study of nearly three million papers published over the course of half a century found that women would not achieve parity for their output in computer science until the year 2100.[349]

The percentage of women researchers decreases with every ascending rung of the academic ladder, from postdoctoral fellow to assistant professor, associate professor, full professor, and up to the president's office.[350]

Statistics Canada (2020) showed that women faculty "have come a long way" since 1970, but that there is still much more work to do. In 1970, women accounted for 14 percent of assistant professors, 8 percent of associate professors, and 3 percent of full professors. In 2018, women reached parity for assistant professors (50 percent), were on the verge of parity for associate professors (44 percent), and made up 29 percent of full professors. While the share of women among full-time faculty is almost at parity in social and behavioral sciences (47 percent), it is lower in agriculture, natural resources, and conservation (31 percent); lower still in physical and life sciences (28 percent); and lowest in architecture, engineering, and related technologies (17 percent).[351]

Almost a decade ago, in 2013, a special section in *Nature* drew attention to the fact that much work remained ahead to achieve gender equality in science: "Science remains institutionally sexist. Despite some progress, women scientists are still paid less, promoted less frequently, win fewer grants, and are more likely to leave research than similarly qualified men."[352] It is disturbing that these words still hold equally true today: globally, the gender gap continues to widen as women progress in their academic careers, with lower participation at each successive rung of the ladder from doctoral student to assistant professor through to full professor.[353]

When I served as Canada's minister of science, women scientists and researchers told me their stories: how many male hiring panels they had faced, the numbers of postdocs and adjunct professorships held by women, and the courses they had taught for a few thousand dollars and no benefits. They were repeatedly asked how serious they were about science, how they would "fit in with the culture of the department," code for not making waves in a largely male department, and whether they might like a family one day. In addition to violating human rights principles at the domestic level, such persistent inequalities are contrary to Article 24 of the UNESCO Recommendation on Science and Scientific Researchers (2017): "States should ensure that scientific researchers enjoy equitable conditions of work, recruitment and promotion, appraisal, training and pay without discrimination."

The problems do not stop here. Science must not only fight sexism but it must also battle racism while confronting its own history: scientists, after all, were responsible for "codifying the concept of race as a biological category" that was both descriptive and hierarchical.[354] The eighteenth-century Swedish botanist Carl Linnaeus, whom generations of high school students have learned about, assigned human beings to such categories as skin color, behavior, and clothing. A nineteenth-century American doctor, Samuel George Morton, documented the supposed differences between Indigenous people and Europeans by looking at their skulls.[355] As briefly mentioned earlier, Francis Galton was the father of eugenics, and his ideas were adopted throughout Europe, Australia, Canada, China, Japan, and the United States. While the immoral, racist and scientifically erroneous theory eventually came under question, it laid the groundwork for forced sterilization laws, the Holocaust, the murder of six million Jews, and other victims of persecution by Nazis and their collaborators, including, Black people, people with disabilities, and LGBTQI people.

Sexism and racism, perpetuated under the guise of science, have enabled and sustained centuries of oppression.[356] Today in the United States, Black people make up roughly 15 percent of the population aged twenty to twenty-four but only 2 percent of those who earn a PhD in physics.[357] As of 2022, there were only twenty-three Black women who had obtained PhDs in Astronomy in the United States and about fifteen Black women who had earned PhDs in Planetary Science.[358] The evidence shows that gender and racial gaps have widened during the COVID-19 pandemic.[359] The entire STEM community has work to do.

* * *

We need more women in science, not only to assert our human rights but also because the full inclusion of women will make science better. Increasing the number of women scientists will bring new perspectives, new research questions, and perhaps new research methodologies. Diversity of perspectives makes science more relevant and accurate: after all, humanity is heterogeneous, not homogeneous, and we better understand people, our relationships with one another, and the world around us when we include women. While it is abundantly clear that there are biological differences between women and men, medicine has been shaped by generations of men who assumed that women were inferior versions of men or simply reproductive bodies. In fact, initial drawings of human skeletons emphasized women's pelvic girdles but under-emphasized the size of women's skulls, lending support to the belief that men were rational and reasoned and that women existed for labor.[360]

The renowned scholar Magnus, writing in the Middle Ages, thought that the "clitoris was homologous to the penis." Sixteenth-century physician Vesalius argued that the clitoris did not appear in "healthy women." The *Malleus Maleficarum* (1486), a standard handbook on witchcraft, including its detection and removal, suggested the clitoris was the "devil's teat," and if found on a woman, "it would prove her status as a witch." In the 1800s, women suffering from "hysteria" were subjected to clitoridectomies.[361]

For centuries, male experts studied men and presumptively extended the results to all people. As recently as 1958, the Baltimore Longitudinal Study, launched to explore "normal human aging," enrolled no women, and then took twenty years to correct the omission.[362] In 1986, researchers at Rockefeller University studied the impact of obesity on estrogen activity and breast and uterine cancer; not one woman was registered in this study.[363] The real-world problem is that half of the planet's population is female and there are clear, sex-based differences in susceptibility to disease, disease prevalence, symptoms, diagnosis, severity, response to treatment, outcomes, and long-term complications.[364]

For decades, women died unnecessarily of heart attacks, which were seen as a man's disease. Women would arrive at the hospital only to be sent home to die of what often looked to their doctors like indigestion. Even today, heart disease among women remains "under-diagnosed, under-treated, and under-researched."[365] Moreover, researchers typically studied white males who were considered "the norm." It was assumed that women would have the same response as men to drugs in clinical trials. This was, in part, because women were known to have fluctuating, pesky hormones that were thought to confound research studies or at least make them more expensive to run.[366] Even the rats and other animals used in scientific experiments were mainly male, as many researchers were afraid that hormone shifts in female animals could affect results.[367]

It was not until 1993 that the U.S. Congress mandated the involvement of women and minorities in clinical trials, and it was in 2014 when the NIH required researchers to start testing female lab animals, female cells and tissues, and to consider sex as a variable in experiment design and analysis.[368] It should come as no surprise, then, that we still "know less about almost every aspect of female biology than we do about male biology."[369]

Caroline Criado Perez, journalist and author of *Invisible Women: Exposing Data Bias in a World Designed for Men*, outlined numerous spectacular failures to take account of women's perspectives. She found "boots that don't fit women's feet," "cell phones that are too big for women's hands," "safety goggles that are too large for women's faces," and "police vests that don't account for women's breasts." Criado Perez also described how the first airbags in cars were designed and tested by mostly male engineers and tested on male prototypes. These devices were installed in cars as early as the 1970s but did not become standard equipment until 1998. Women and children, who are, on average, smaller than men, were injured and killed. It was not until 2011 that the U.S. Government tested the impact of a crash on a belt-restrained female prototype.[370] Throughout her book, Criado Perez made the case that men are not the norm, "women are not outliers," and we must design and build for all humanity. Also, for products to be safe and effective, they need to be tested on women.[371]

"There is an urgent need for more women to participate in and lead the design, development, and deployment of artificial intelligence (AI) systems."[372] A critical absence of women and women-led perspectives in AI—driver of the next industrial revolution—may threaten the notion of "AI for good." The voices of women and girls must also be included in any discussions of the risks of AI, including their online safety. AI already has a gender and race problem that "reflects the priorities, values, and limitations of those who hold power."[373] Because gender and racial biases are found in data and training sets, devices and algorithms have the potential to reinforce and spread harmful stereotypes and further marginalize and stigmatize underrepresented groups.[374] To date, there have been many examples of AI systems discriminating against specific populations. For example, when news articles written in one language are converted to English, phrases referring to women often become "he said" or "he wrote." Software designed to warn a photographer when a subject is blinking tends to interpret Asian subjects as always blinking.[375] And few image databases contain details on ethnicity or skin type, with the result that AI skin cancer diagnoses risk being less accurate for dark skin.[376] We are at risk of amplifying these biases in society and perpetuating harmful stereotypes through AI.[377] We cannot afford to build into our collective future the same exclusionary gender and race bias that has persisted for centuries and held us all back.

These intersectional gaps are particularly worrying. Diversity is not only a social good but also key to research excellence, innovation, and economic growth; a lack of diversity means a loss of talent. Only 3 percent of Black women, 4 percent of Latinas, and 5 percent of Asian women earn STEM degrees in the United States.[378] Half of all babies born in America today are girls, and half of all babies are "racial or ethnic minorities."[379]

* * *

There is also an economic imperative for women in science. Today, the highest-growth careers are in STEM and the demand for STEM skills will continue to increase. For example, in the European market, demand for STEM skills is predicted to rise from 8 percent to 23 percent between 2015 and 2025; and demand for STEM jobs is predicted to increase by 7 percent.[380] Data from the European Union show that increasing the participation of women in STEM subjects would have a strong positive impact on GDP. Closing the gender gap could increase GDP by €610 billion to €820 billion and increase per capita GDP by 2.2 percent to 3.0 percent in 2050.[381] The McKinsey Global Institute showed that closing the gender gap could add nearly $28 trillion to global GDP, or an increase nearly the size of the American and Chinese economies combined.[382] Despite this economic opportunity and the expanding need for a specialized workforce, not everyone has equal opportunity to succeed in STEM careers.[383] Although there are innumerable outstanding, talented women across the globe, their progress has been relatively slow over the last five decades.

While women have achieved global parity at the bachelor's and master's levels of study across disciplines and are at the tipping point of gender parity at the doctoral level, only one in five countries has achieved gender parity at the career level.[384] In only a handful of countries do women working in science outnumber men.[385] There has been an attempt at gender parity for STEM researchers in Central Asia, Southeast Europe, Latin America, and the Caribbean. "These regions are home to ten of the top twenty countries for their share of women researchers": namely "Venezuela (61 percent), Trinidad and Tobago (56 percent), Argentina (54 percent), North Macedonia and Kazakhstan (53 percent), Serbia (51 percent), Montenegro (50 percent), and Cuba, Paraguay, and Uruguay (49 percent)." "The consistent high ratio of women researchers in many European and Asian countries is in

part a legacy of the Soviet Union, which valued gender equality."
This is true, for example, of "Azerbaijan (59 percent), Georgia
and Kazakhstan (53 percent), Serbia (51 percent), and Armenia
(50 percent)."[386] Between 1962 and 1964, 40 percent of chemistry
PhDs were awarded to women in Soviet Russia compared to only
5 percent of chemistry PhDs awarded to American women at
that time. In 2013, 37 percent of chemistry PhDs were awarded
to American women.[387] Across sixty-eight countries today, women
make up about 40 percent of the STEM workforce.[388]

Even when women earn STEM degrees, many do not pursue or
stay in STEM occupations.[389] And those women who do enter the
STEM workforce are more likely than men to leave STEM careers.[390]
"Persistence" in STEM—the proportion of women (or men) in each
field who are still in that field after several years—is already lower
for women than men. In 2010, 66 percent of women remained in
STEM after the first year, compared to 72 percent of men.[391]

A longitudinal study from Canada showed that between 2006 and
2016, 34 percent of women with STEM credentials moved on to non-
STEM jobs, compared to 26 percent of their male counterparts.[392]
While Canadian women earned 34 percent of STEM bachelor's
degrees, they accounted for 23 percent of science and technology
workers aged twenty-five to sixty-four.[393] Men make up 77 percent
of the workforce. In 2006 and 2016, half of male STEM graduates
worked in non-STEM occupations while over six in ten women with
STEM credentials worked in non-STEM occupations. A higher
proportion of men had managerial occupations in both 2006 and
2016, and women with STEM credentials were more likely to be
unemployed than their male counterparts.[394]

In the United States, women made up about 47 percent of the
total workforce, but about 27 percent of STEM workers in 2019.

In other words, American men make up almost three-quarters of the STEM workforce.[395]

The deeper you drill into the sector-by-sector data, the more discouraging the results. Women are well-represented in health, for instance, but are severely under-represented in engineering, accounting for only 28 percent of engineering graduates globally. Percentages are lower when it comes to individual members of the Organization for Economic Co-operation and Development (OECD): Canada (20 percent), Chile (18 percent), Ireland (18 percent), Japan (14 percent), Switzerland (16 percent), the United Kingdom (17 percent), and the United States of America (20 percent).[396] Women account for 15 percent of engineering occupations and about 25 percent of computer workers—the latter of which declined between 1990 and 2019.[397]

Importantly, "a 2017 study of over 5,000 women who earned bachelor's degrees in engineering found that 10 percent never entered the field and 27 percent left the profession." Of the women who choose to leave the engineering profession, 30 percent cited workplace environment as the reason.[398] Familiar reasons are typically behind the decision to leave, including inequitable compensation, poor working conditions, inflexible work environments, lack of recognition at work, and inadequate opportunities for advancement.[399] The problem is the workplace, not the women.

Data from 2018 to 2019 show that men continue to dominate the tech sector and its leadership. Facebook's technical roles include 23 percent women and its leadership roles include 33 percent women. Apple's technical roles include 23 percent women while leadership roles are 29 percent women. Google and Microsoft had similar results.[400]

The most relevant fields to driving the future economy—computing, engineering, IT, mathematics, and physics—are those in which women remain underrepresented: cloud computing (12 percent), engineering (15 percent), and data and AI (26 percent).[401]

The AI sector is expanding rapidly.[402] In fact, the number of workers with AI skills increased by 190 percent between 2015 and 2017.[403] Educated and skilled women remain under-represented in this field. Globally, only one in five workers in AI is a woman.[404] Within the top twenty countries for AI, the "share of women professionals with AI skills" ranges from 14 percent for Mexico to 28 percent for Italy, South Africa, and Singapore.[405] Women suffer discrimination, lower hiring rates, under-representation, and limited professional opportunities in AI. According to Dr. Radhika Dirks, CEO and cofounder of Ribo AI, "People tend to pattern match to what has been successful in the past, and leaders historically have looked a certain way."[406]

In the EU, more than half of men earning degrees in IT worked in digital jobs, compared to just one-quarter of women.[407] According to a 2019 Silicon Valley Bank study of start-ups in technology and healthcare in Canada, China, the United Kingdom, and the United States, only 28 percent of them had a woman among their founders, and almost half (46 percent) had no women at all in executive positions.[408] Even when women lead start-ups in tech fields, they struggle to access venture capital and other forms of financial support. A 2020 global survey of 700 firms by TrustRadius showed that a mere 2 percent of venture capital is being directed toward start-ups led by women.[409] In 2021, in response to the financial challenges that women entrepreneurs face, the Canadian government developed the first-ever Women Entrepreneurship Strategy—a $7 billion initiative to remove systemic barriers and access financing, resources, and networks. In the U.S. Congress, lawmakers are now looking to hold venture capital firms accountable for fostering the "tech bro culture" as data shows that most companies that "firms invest in are owned by white men."[410]

* * *

While the underrepresentation of women in science was long recognized, for much of my academic career there were two very different conversations about women scientists. The predominant conversation occurred when men were the majority present in the room. It focused on remembering that "academia is first and foremost about excellence," but was also quick to acknowledge "just how far women had come in science." In pulling apart that message, we discover that, because academia is about excellence, we must protect the status quo to maintain excellence. The fact that we have fewer women, under this narrative, is not a problem because we must prevent the dilution of excellence. The underlying message is that somehow "women are less excellent," and for far too long, women's voices in science were seen as unnecessary and unwanted. It is familiar, tired, and out of touch with the realities that women scientists face even today.

The second conversation, equally revealing, took place when men were not present. Women spoke about their hurt: the microaggressions, the under-attention, under-employment, under-funding, and under-payment they had experienced. They talked about how they wanted both a family and an academic career or how they wanted to talk out against a colleague who had harassed someone but felt silenced because he had tenure or because "that's just the way things are." I know these challenges personally. When I was a tenured professor, a fellow faculty member shot a question at me during a staff meeting: when did I plan on getting pregnant? On other occasions, I was asked how I wanted to be treated: as a woman or as a scientist. One university official made me very uncomfortable while he explained to me: "It must have been hard for you. Pretty girls finish last in science. It's not like business where good looks, six-foot-two, and a suit of blue gets you the job." Later, when I asked my university president why I was being paid in the bottom tenth percentile, I was told point blank that it was because I was a woman.

As a result of these experiences and others like them, I spent my academic career fighting for equality, diversity, inclusion, and belonging in research. Like so many women scientists, I learned to battle sharp elbows and sexism when I led my 1998 research expedition to the Arctic. I insisted that the university press that published my book on this expedition keep references to the hardships of being a woman scientist and, particularly, an early career researcher. Later, I hired ten young women as TAs at one time so that they would know they had a champion. I pored over data, learned the research, and developed seminars on how to increase women's participation and inclusion at the university. While it is impossible to shed light on all the challenges women scientists face at the university, the next sections highlight some key areas.

In the past, doctoral students who had finished their degree could progress directly to an academic career. This expectation has changed significantly: today, postdoctoral training is required for almost every academic discipline, with the added benefit of providing cheap research and teaching labor for the institution. The expectations of postdocs depend upon the country, institutions, and career paths they choose to pursue. The pursuit of a career in academia involves research, funding searches, publishing in high-impact journals, and teaching. The selection process for these positions is fiercely competitive and men continue to win more postdocs than women. Centuries of prejudice and the continuing stereotype of the male genius, especially the white male genius, remain entrenched, despite the assertion by academic institutions that they espouse progressive ideas and create and disseminate new knowledge.

Overall, "the scientific enterprise remains deeply conservative".[411] Laboratories that have a male principal investigator represent over 70 percent of all labs in the United States, and male faculty employ 20 percent fewer female postdoctoral researchers than female faculty. Moreover, the more decorated a male scientist is, the fewer women he trains, and universities tend to hire their junior faculty from elite men's labs.[412] The bottom line is that a reduced number of women postdocs leads to a smaller talent pool of women who are available and willing to compete for faculty jobs.[413]

Several forms of bias are responsible. Affinity bias: a preference for people who are like us. Confirmation bias: a preference for information that supports an evaluator's initial assessment. And availability bias: a preference for information they have viewed recently.

Most significantly, the misplaced concern that women may leave to have a family continues to rear its ugly head. If a woman scientist who wants a family starts her postdoc at thirty-four or thirty-five years of age and then takes on a three-to-five-year postdoc, when does she choose to have a baby? How does she balance fertility, long days, low salaries, and high costs of early learning and childcare?

A survey of L'Oréal USA For Women in Science (FWIS) Fellows found that 90 percent of respondents agree that "women's opportunities for career advancement in science have improved in the last decade;" however, a similar percentage of respondents reported that "gender discrimination remains a career obstacle." The same survey also found that 89 percent and 73 percent of respondents reported that gender bias and sexual harassment, respectively, serve as obstacles to women's career trajectories, specifically in the post-doctoral stage. Moreover, only 45 percent of respondents thought that "women entering their specific field of study are given equal opportunities to men to pursue their careers."[414]

With the postdoc finished, many women in STEM fields will look to land a coveted tenure-track position at a university, for which they will need to compete with hundreds of qualified applicants.

Ultimately, tenure grants a professor permanent employment, protects her, him, or them from being fired without cause, and allows research and teaching on any topic. It is an earned status that is closely linked to academic freedom. However, the number of tenure-track positions has decreased as universities increasingly use cheaper adjunct, contract labor, and the number of doctoral degrees awarded continues to climb. In Canada, tenure-track positions decreased by 15 percent between 2009 and 2017.[415] Moreover, there is little consistency among universities and departments about what an application and interview for tenure track entails. As a baseline, a candidate can expect to submit documentation, including curriculum vitae (CV), statement of research, letters of recommendation, and potential writing samples; an outline of her teaching philosophy, style, or past courses; and past teaching evaluations.[416]

A woman's CV captures with whom she has worked, whether she has spent time in an elite lab or at a prestigious institution, and what she has published. It also reflects baked-in biases that she encountered during the training process, which can profoundly impact her prospects.[417] If she is fortunate to be shortlisted based on her background, she will need to prepare for an intense interview process. Perhaps a video interview, one or two visits to the institution, each of which may be two days long, and numerous meetings with university officials, search committee members, and faculty. She will need to be impressive throughout the process, show her expertise, and demonstrate that she can indeed cut it in the academic scientific world. Far too often, her successes will be discounted or ignored. While there is an ever-growing business case for diversity, what happens if she is interviewing at a male-dominated department or, in some cases, an all-male department, or a university where the leadership is predominantly male and white? Again, people tend to have a bias in favor of individuals like them. Gender bias, often concerned with preserving the status quo, is pervasive amongst female and male faculty.[418]

In the United States, the American Association of University Women reports that only 15 percent of tenure-track engineering faculty and only 14 percent of tenure-track computer science faculty are women. Grogan (2019) reported that "even in disciplines in which female and male faculty are hired at rates proportional to the gendered PhD graduation rates, women are significantly more likely to become faculty in non-PhD-granting institutions, whereas men are more likely to become faculty at PhD-granting institutions." To be relevant and ranked highly in various surveys, academic institutions must show candidates and, for that matter, the public, that they welcome people from all backgrounds. In contrast to what was common in the past—a narrowly defined post that only one white man in the country could possibly fill—universities today must define job positions in a gender-inclusive way and as broadly as reasonable to attract the largest pool of highly qualified candidates. To facilitate this work, the hiring department should appoint a gender-balanced, diverse search committee that undergoes anti-discrimination and implicit bias training before beginning the actual hiring process.

If a woman receives a job offer, the next step is salary negotiation. I tell women scientists who are offered their first job to first research the salary range, benefits, and the research start-up package for their discipline at similar universities. I tell them to ask for the comparators in the department and the university, at least until the day these figures are made public. Better yet, they should try to obtain inside knowledge because if they are lowballed on salary, it will not only impact their monthly and annual pay but it will also haunt their future academic career and affect their pension. The starting point matters profoundly: it remains the foundation of much inequality and unfairness.[419] Numerous reports show that women continue to be paid less than men. In 2019–2020 in Canada, for example, the pay gap remains pervasive across Canadian universities. The percentage difference in average salary for full-time teaching staff at select large research universities can be as high as 15 percent.[420]

The gender pay gap can result from "bias in determining starting salaries and subsequent merit pay, from differing rates of promotion, and from the impacts of parental and caregiving leave." Recent research shows that over the course of a woman professor's career and into retirement, this pay and pension gap leads to a difference of roughly half-a-million dollars. Unfortunately, this research does not include race data. But the Canadian Association of University Teachers has data that shows racialized professors experience a "10 percent pay gap relative to their non-racialized peers" and racialized women professors experience even greater inequity.[421] Incidentally, the only reason we have much of this data today is because as minister I reinstated an important academic survey. Unfortunately, universities have yet to agree on releasing data for contract academic staff, wherein many early career researchers, women, and diverse scholars, teach courses for several thousand dollars each, often with little to no benefits.

When I was twenty-four years old, I interviewed for tenure-track positions at two different universities. One university administration authoritatively advised me that "we start people like you at $38,000." I have always wondered what "people like you" meant. A second university told me that "the salary is $48,000." Both were fixed salaries. No data were presented and no sliding scales were shown, just two numbers, one significantly higher than the other. Because I would be one of only two women in the department, I had no women role models to whom I could ask questions. I did not know how much men were getting paid. In short, I was not in a strong position to negotiate. A few years into the job, I started to interview at other universities and I was surprised when I was consistently shown salaries at least $20,000 higher than my salary as a tenured associate professor. I called my own faculty association to inquire about my salary and learned that I was being paid in the bottom tenth percentile of the university and that there were men being hired at a higher salary than me who had yet to finish their PhDs. I made an appointment with the president and learned that I had two strikes against me. I was a woman and I was young, but I really shouldn't worry as there were "other women in the same boat." When I countered that there were men being hired at a higher salary with less qualifications, I was told they were older and had a family. Colleagues thought this was acceptable and not the responsibility of the university to advocate for half their student population with employers.

Administrators have a choice to recognize bias by looking at the data[422] and paying women, men, gender-diverse people, and other underrepresented groups equally. To fix the problem, universities have taken an incremental approach. Slowly over time and under pressure, they have increased women's salaries. Universities have also put in place band-aid solutions, such as salary anomaly studies and wage adjustments,[423] rather than addressing the underlying biases that caused the inequities in the first place. If Canada could double the number of women employed during the Second World War, why can't a university ensure a few thousand faculty are paid equally? Since 2015, multiple Canadian universities have made pay adjustments for women.[424] I, too, have been the recipient of such an "equity adjustment." A one-time payment of less than $1,000 does not make up for gross underpayment for many years and pension contributions.

These gaps also exist in the STEM field outside the academy. While women earn higher wages in STEM occupations relative to most other jobs, they still earn less than men. Women with STEM bachelor's degrees, if they stay in STEM fields, are more likely to be channeled into lower-paying technical roles within companies, as opposed to professional roles.[425] A survey of almost 600 computer science and engineering students who graduated from more than two dozen American institutions between 2015 and 2017 found that women, on average, earned less than $61,000 in their first jobs compared to over $65,000 for men, despite having the same degrees and grade point averages.[426] Starting out $4,000 behind men makes it almost impossible to catch up.[427] Another American study found that 75 percent of the technology occupations examined had gender pay gaps above the U.S. national average of almost five-and-a-half percent.[428] Moreover, almost one-third of women in STEM (29 percent) report earning less than a man doing the same job; only 6 percent of men report earning less than a woman.[429]

One of the main reasons why the gender pay gap persists today is that it remains hidden from policymakers. Statistics Canada (2019) studied pay transparency and the pay gap among university faculty. It found public sector salary disclosure laws "reduced the gender wage gap by approximately 2.2 to 2.4 percentage points, representing a 30 percent reduction in the pay gap."[430]

Men and women must continue to expose these hidden gaps. It's time to stop blaming women for not negotiating and to publicize compensation, expose the disparities, and fix them. Erasing financial disparities is a crucial step in promoting the equal advancement of all women, but it is only one step. Others must follow.

To reach tenure, a woman scientist will need to go through an elaborate tenure review process, which can be daunting, demanding, and nerve-wracking. She will need to produce a record of her publications, teaching records and evaluations, university service, letters from the department, and outside reference letters assessing her scholarship and standing in the field. Next, there is likely a departmental vote and if there is a decision to recommend tenure, her dossier will then be passed to a committee of tenured faculty drawn from a range of departments who may or may not approve the departmental recommendation. Usually, the president or senior administrator makes the ultimate decision and may choose to reject unanimous recommendations.

In 2019, Statistics Canada reviewed data from the University and College Academic Staff System (UCASS) survey, which I reinstated as minister (it is an important driver of evidence-based decision-making), and the Survey of Postsecondary Faculty and Researchers to determine gender differences in tenure among Canadian university faculty. This study found that women were less likely than men to feel that hiring and promotions were fair and equitable decisions at their institutions, "even after considering age, education, instructional program, and other characteristics." Racialized researchers, individuals with disabilities, and "sexual minorities" were also more likely to state that hiring and promotions at their institution were inequitable, and women with a disability were twice as likely as women without a disability to feel that practices were unfair.[431]

For a long time, it was believed that every woman who became a tenured university professor had to take a straight journey through an academic pipeline, earning a bachelor's degree, a master's, and a PhD before moving up from assistant to associate to full professor. That's not always the case. By the time young women finish secondary school, some women may already be a distance along their STEM journeys. Others may just be starting out, and still others may not start for years to come. There are multiple paths, and they are not always easy or simple. They can meander, branch off, stop, and start again. I know one inspiring PhD student who was raising three children when she decided to finish high school and study part-time to earn a bachelor's, master's, and doctoral degree while teaching hundreds of university students. Regardless of the individual journey, it is always women choosing their own paths, determining their destinations, and finding ways around and over layers of obstacles. After all, multiple studies show that at nearly every step of development, women experience marginalization in the classroom and in the workplace.[432]

To achieve tenure, a woman scientist must meaningfully contribute to her research field, contribute to the university and the broader community, and teach. She must secure research funding and work to become a principal investigator. Women are less likely to receive tenure and 10 percent to 20 percent less likely to become independent principal investigators (PIs) in research labs than men.[433] In 2019, 63 percent of tenured positions in Canadian universities were held by men compared to 37 percent by women—up substantially from 1990, when only 14 percent of tenured professors were women.[434]

Male PhDs reported an average of U.S. $936,000 in start-up funds, whereas female PhDs reported U.S. $348,000, reflecting a difference of almost $600,000.[435] The prestige and value of each grant matter, yet a typical NIH research grant to a male principal investigator is $41,000 larger than a similar grant to a woman, and the gap is even larger at top universities: $76,500 at Brown and $68,800 at Yale.[436] Moreover, women submit fewer grant applications than men to the NIH (31 percent versus 69 percent, respectively) and National Science Foundation (24 percent and 76 percent, respectively). Women hold fewer NIH grants, submit fewer renewal grants, and are less likely to have them succeed.[437]

Like elsewhere, race in the sciences leads to systemic disadvantage and exclusion. "After controlling for education, training, and publications, African American NIH applicants are 10 percent less likely to be awarded a grant than white investigators."[438] Career advancement also depends on scientific productivity, which is measured in the number of articles published, preferably in high-impact journals, and the number of citations. Studies across many STEM fields have documented that women author fewer papers than expected based on the percentage of women in the field. But let's not forget that they are also less likely to be credited with authorship. A new study looked at nearly 10,000 U.S.-based scientific teams, including almost 130,000 team members, that published nearly 40,000 articles between 2013 and 2016. The authors compared who actually did the work with who got credit for it. They found that women are less likely to be authors than men in their own research groups as every PI, the majority of whom are male, assigns authorship according to personal standards and bias can creep in.[439] "Papers in high-impact journals with men in key author positions are then cited more frequently (thirty-nine citations per paper) than papers with women in key author positions in those same journals (thirty-five citations per paper). Women are also significantly under-represented on journal editorial boards and as journal reviewers."[440]

Women scientists face inequitable hurdles that result from pregnancy, breastfeeding, caretaking, prejudice, and socially driven childcare demands.[441] As a result, many women talk about paying a "baby penalty," a heart-breaking term for welcoming a child into the world. Research by Cech et al. (2019) showed that both new mothers and fathers were more likely to leave STEM occupations than "childless peers," and most did not return to these occupations when their children reached school age.[442] Moreover, researchers in the United States found that "42 percent of women left full-time STEM occupations within three years of having their first child, compared with 15 percent of new fathers."[443] Women who had children "while they were graduate students or postdoctoral fellows were more than twice as likely as new fathers or single women to turn away from an academic research career." Moreover, single, childless women attained "their first tenure-track jobs at higher rates than wives, mothers, or single men and almost at the same rate as married fathers."[444] The penalties are harsher for racialized women.

While having a family can be a slight career advantage for men, it can often be a "career killer" for women.[445] One male academic put it to me this way: "Having kids made me a hero, but for my wife, it made getting tenure-track positions next to impossible and the backlash from her supervisor was awful. He even warned her about having another child. It's not easy, but we're doing our best to figure it out."

A 2021 research study looked at the impact of parenthood on scholarship. The authors analyzed parental leave policies, timing of parenthood, publication data, and perceptions of research expectations among 3,000 tenure-track faculty at 450 PhD-granting computer science, history, and business departments in the United States and Canada. Parenthood explained "most of the gender productivity gap," with the penalty being strongest in the years right after parenthood. The penalty was found to be shrinking over time, likely due to better policies.[446]

When I was minister, many women told me that they ran into tired thinking—perhaps more accurately, a lack of thinking—in fossilized, patriarchal STEM cultures. If they had a child or other responsibilities outside of work, they weren't seen as committed, dedicated, or serious scientists. This needs to change. Women should not have to apologize or be seen as less devoted to science for giving life and taking on caregiving obligations. I asked for the parental benefits policies for graduate students and postdocs at the three federal granting councils. Two councils gave six months of parental leave and one council gave only four months. Yet the law in Canada requires twelve months of paid parental leave. I changed the policy and the funding. Today, each grant recipient will receive twelve months of paid parental leave. I hope this action will at last put an end to the long-standing question for women researchers in Canada: "Do I pursue an academic career or a family?" These goals must be entirely compatible.

More work needs to be done to accommodate women who want families:[447] flexible workplaces, flexible career tracks, a reentry policy, and childcare assistance.[448] In the United States, due to a "lack of affordable, high-quality childcare, nearly half of the female scientists leave full-time science after their first child is born." This hurts the sciences and all of society.[449] In 2021, Canada committed to building a nationwide early learning and childcare system with provinces and territories. We also need to look at grants to keep women-run laboratories going while women researchers take maternity leave and childcare grants so that women will be able to attend conferences with accommodations such as babywearing, changing stations, and nursing areas to present their research.[450]

The COVID-19 pandemic, a time of tears, trials, and tragedy, magnified long-standing gender inequities, creating the so-called "she-cession" and bringing the plight of women caregivers to light.[451] The pandemic also led to the closure of universities and labs. Women scientists, technicians, and healthcare workers— caring for children, seniors, people with disabilities, and other family members—were the most affected.[452] Women scientists told me that their care responsibilities increased, and their productivity decreased, making funding applications, publishing, and achieving tenure even harder. These impacts will persist for many years to come.[453] We must do everything possible to prevent and reverse the erosion of hard-fought gains.

Career advancement also depends on visibility, reputation, and awards, but bias plays a role here, too. Women can experience a collaboration bias: men are 15 percent more likely to "share data when the request comes from another man."[454] Sometimes, there has been no request, rather the data was just taken, as it was with Dr. Rosalind Franklin. I was fortunate enough to have been tipped off by the British media that American and British male team members of the Svalbard expedition had conspired to release the results and claim credit for my work and that of other team members. Throughout history, women scientists have fallen victim to the so-called Matilda Effect—the denial or repression of the contributions of female researchers to science.[455] In 1915, Alice Ball graduated from the University of Hawaii, becoming the first woman and first Black master's degree holder, as well as the first Black woman chemistry professor. Within a year, she had developed a treatment for Hansen's disease, more commonly known as leprosy, that often led to exclusion and exile. But after Ball died at the age of twenty-four in 1916, a male colleague claimed her work as his own.[456]

In the late 1960s, PhD student Jocelyn Bell Burnell discovered pulsars, spinning neutron stars with a mass greater than the sun, but the corresponding 1974 Nobel Prize in Physics went to her male supervisor.[457] As the "girl researcher" and an outsider, she experienced impostor syndrome. She doubted her own achievements and feared that she would be exposed as a fraud. The media questioned if she was blonde or brunette and the size of her bust.[458] In their eyes, she barely rated as a scientist. Forty years after she was overlooked for the Nobel, Dame Jocelyn Bell Burnell was awarded science's most lucrative award: the Breakthrough Prize. She donated her winnings toward PhD studentships for people from underrepresented groups in physics.[459]

Research shows that men are invited to speak on scientific panels twice as often as women and are more likely to be colloquium speakers at prestigious universities.[460] However, the "presence of women colloquium chairs increases the likelihood of a woman speaker," another reason why the inclusion of women in leadership roles matters profoundly. Women account for less than 25 percent of recipients of the most prominent awards in some STEM fields and are notoriously underrepresented among Nobel winners in science.[461] By 2021, twelve of the two hundred twenty-four Nobel laureates for physiology or medicine were women. Seven of the one hundred eighty-eight Nobel laureates for chemistry were women, and only four of the two hundred nineteen Nobel laureates for physics were women.[462] Women account for just 3 percent of science category winners and they were almost always jointly awarded with male peers.[463]

In 2018, Dr. Donna Strickland became only the third woman over a period of 117 years to be awarded a Nobel Prize in physics. Her tremendous achievement inspired people all over the world. Soon after, it was revealed that she was an associate professor and had not been promoted to the highest rank of full professor, which would come with a salary increase and other benefits.[464] Women professors quipped that, "Men expect to make full professor; women need a Nobel Prize and then maybe?" While women have come a long way and much has improved over centuries, there is still a long way to go and an ongoing fight for recognition.

I have heard from innumerable women how they felt that their research was underappreciated and that they had to fight constantly to maintain their place and influence within the department or at a conference, despite having tenure and profile. I share their experience. I also know firsthand what it is like to be labeled "woman" or worse yet, "young lady" in these roles. Like many women scientists thirty years ago, I was labeled "passionate," with a "breathy, emotion-laden intensity," and "unserious." After all, only male scientists could be serious scientists. Never was it acknowledged that it was a woman who had pulled together the world's best and forced them to work together. The blatant sexism and sharp elbows were not acknowledged, nor some of the worst personal attacks that came from women journalists and scientists themselves. After a less-than-balanced documentary aired on the Svalbard expedition, one member of our department who was also a high-ranking official of the university walked into the faculty lounge and exclaimed, "So Duncan finally got what's hers." In his esteemed opinion, I had at last been knocked off my perch.

I was thirty-two years old, easily twenty years his junior, and yet this senior academic felt the need to disparage me in my absence and in front of faculty colleagues during breakfast. Each chosen or thoughtless action, behavior, or comment leaves a mark, erodes confidence, and undermines women; and for many women, those scars remain close to the surface and last a lifetime. A mid-career scientist once shared, "I finally get it—it wasn't me. I wasn't wrong. They were wrong; the system was wrong. I can at last let some of the hurt go that I have been carrying for far too long."

No caring father, uncle, or brother would choose to do this to his daughter, niece, or sister. Yet often, the very same men do this to women scientists. Women scientists have had to be resilient in a way that is not required of men, whose gender poses no systemic obstacle and continues to generate opportunity on its own. Still, the career slams add up. Women drop out and leave STEM fields. It wasn't long after the faculty lounge incident that I decided to leave the university. I was thirty-three years of age. I had tenure. I was an associate professor. I had led a major international scientific expedition. I was writing a book and the Canadian government had asked me to serve on the Intergovernmental Panel on Climate Change. To those on the outside, it looked like I had a good career and a bright future in academia. But I was tired of being underpaid, tired of having to put a door between a colleague and me every time he wanted to "talk," and tired of fighting to be treated equally. I didn't know what I would do, but I knew it was time to leave and that I would figure it out in time. And no matter what I chose to do, I knew I would focus on women, empowering them, and making sure that they never faced the challenges I had.

I entered politics, ran for Parliament, and was appointed Canada's minister of science. The reporter Maryn McKenna, who had covered the Svalbard expedition in the 1990s, wrote at the time of my appointment that I was a "badass" and that the Norwegian excursion had been bold and imaginative. It had fired the public imagination. If I could "marshal government support and public opinion in the same way" that I had "muscled a dozen male scientists twice my age" to the Norwegian Arctic, then "bruised, muzzled Canadian science may have found the defender it needed."[465]

Unfortunately, that was not always the response I received from the political or scientific community in Canada. Some members of the scientific community were accustomed to male ministers, which turned out to be important, even if my male predecessor did not want to confirm whether he believed in evolution. I found it hard to believe that two-centuries-old thinking still existed in the corridors of power and academe, but I was seen by some as a woman who had leapt from the wrong academic discipline to the wrong political portfolio. My research and teaching were dismissed, as was my service on the Intergovernmental Panel on Climate Change. I could not possibly understand the complexities and needs of the academic and scientific communities. I had academics lecture me on what master's and doctoral degrees were, in case I could not grasp the concept of academic higher degrees. And more than once, male scientists explained to me why women were not as successful as men in science and how men innately understand the scientific system better.

As minister of science, I had four overarching goals: to end the war on science in Canada; to return science to its rightful place in government decision-making; to adequately fund research; and to ensure that science could never be undermined again.

Under the previous administration of Prime Minister Stephen Harper, government scientists had been muzzled, science funding cut, scientific evidence ignored, and world-class facilities such as the Experimental Lakes Area were under attack and facing closure.[466] Science in Canada was attracting international attention for all the wrong reasons. Two thousand scientists and concerned citizens held a mock funeral, complete with a wooden coffin and scientists dressed in white lab coats and black mourning clothes, to lament the death of evidence on Parliament Hill in Ottawa.[467]

For four years, I worked hard to back my goals with actions. I encouraged government scientists to speak out and worked with colleagues to make changes so they could freely express themselves about their work. I undertook the first review of science funding in forty years and secured the largest investment in research in our country's history. I created the position of chief science advisor, staffed it with a woman scientist, and asked her to build a network of chief scientists to provide advice to the prime minister and cabinet. I also made changes to ensure that scientific evidence was considered at the cabinet table. I was equally determined to tackle the pesky problem of equity, diversity, inclusion, and belonging in science. My own ministerial team and innumerable other stakeholders did their best to talk me out of it, cautioning me that it would be "political suicide." I knew firsthand the numerous challenges that women scientists faced. I had lived them, I had witnessed what colleagues and friends faced, often daily, and I refused to be part of maintaining the status quo. I told my team: "You can't expect me to be in this position and not take action."

I began by asking for what any scientist would ask for: data. I wanted data on under-represented groups applying for grants and their success rates, representation on grant adjudication panels, abuse, discrimination, and harassment data, information on pay, and more. I drilled down on the data as hard as I could and when the data I asked for didn't exist, I insisted it be collected so that we could measure progress over time in a consistent, reliable fashion. When I looked at the evidence, I realized that while it had taken 200 years for 13 percent of licensed engineers in Canada to be women, the overall percentage of women employed in the country increased by a staggering 100 percent in just six years during the Second World War.[468] Women who had been raising families were suddenly learning to drive trucks, manufacturing munitions, repairing airplanes, and working as laboratory technicians on the home front.[469] But at the end of the war, these same women were unceremoniously cast aside, incentives such as free government nurseries were withdrawn and women's participation in the workforce plummeted.[470] The lesson until then: if men wanted more women to work, there would be more women, and quickly. If men wanted women gone, they would be gone, equally quickly.

In addition to collecting data, I listened to lived experiences. I traveled across the country visiting university and college campuses, hospitals, and important research facilities. I heard first-hand stories of discrimination, harassment, neglect, and people leaving science. Brave women repeatedly told me that they struggle with the choice between an academic career and having a baby. One woman told me about the months she spent wearing a large lab coat to hide her pregnancy for fear of losing her job. Today, it should be abundantly clear that women are able to have both if they want both and it's our job to finally put in place the necessary policies to allow people to be parents and to succeed in science. Another woman drew a normal bell curve and asked, "Who becomes an astronaut?" She repeatedly pointed to the top of the bell curve: "he is a man, he is most often white, he is usually from Europe or the United States, he usually speaks English or Russian." She then drew a line away from the very top of the curve down to the bottom and said, "What if you are a woman? What if you are not from Europe or the United States?" She asked, "Have I made my point? Or should I continue?" Another person said, "I am Black, gay, and I have a disability . . . Do you have any idea how hard my path has been?"

Indigenous researchers talked about tokenism and cultural extraction, as opposed to knowledge sharing and exchange. To do my part, I launched the first science and research dialogue with First Nations, Inuit and Metis researchers and a new research program.

Researchers with disabilities were not only excluded from the classroom, conferences, and labs but also from discussions of diversity and inclusion. More than one early career researcher told me that he left science because he was tired of hearing that gays don't belong in hard sciences.

Discrimination and bigotry reduce the talent pool of the scientific workforce, poison science culture, and make us rightly uncomfortable. We must get out of the mindset that the change that has been slowly happening over the past decades is somehow acceptable because change simply takes time. There are countless people who have been knocking on the door and hoping to be let in for a very long time. Tellingly, even when new voices are finally given a seat at the table, their treatment has not always been warm and welcoming. Millennia-long inequalities will not change simply because we graduate more graduate students or more PhDs. Patriarchy continues and colonialism and slavery cast their long, awful shadows.

After I had seen the data and heard lived experiences, it became very clear to me that incremental change cannot be the acceptable path to progress in science, not in a rapidly changing world where everyone's contribution is urgently needed. The statistics are stubbornly persistent; they reflect a bias that has existed for hundreds of years. Someone needed to "light a fire." Unfortunately, I was up against many stakeholders who were happy with "the way things have always been done" and who were starved for funds because of Conservative cuts. All they were interested in was money and lab space. It was a challenging environment in which to fight for long-taboo inclusion in research. But if not then, when?

To address under-representation in science in Canada, I began with our prestigious Canada Research Chairs program, itself steeped in patriarchy and academic privilege. For two decades, there had been warnings about the inequity in the chairs program, without any consequences. I gave our universities two years to meet their mandated targets for women scholars, Indigenous investigators, racialized researchers, and academics with disabilities. I threatened to withhold funding if they failed. Not only did the universities make the targets, they surpassed them. I continue to hear compelling evidence of how these new voices are asking new research questions and making science better.

That was only one research program. I needed to start a national discussion about representation in science. It was not easy. There was pushback and backlash and, quite frankly, I often felt that there was a target on my back. On one particularly tough day, one of the young women who worked with me said, "I want to tell you a story that happened to me a few years ago. My professor kept emailing me, he wanted to take me to dinner; he made me feel uncomfortable. One day I had to go to his office to get my final mark." She paused, "Nothing bad happened, but he made me twirl to get my grade. I did nothing wrong, but he made me feel dirty." I got past my tough day and I worked with the civil service to put in place specific, achievable targets across research programs. We agreed that the goal could not just be counting people. In my mind, it had become fundamentally important to address science culture and environment, which is why I worked with the research community across the country. Together we created the Dimensions program, based on the United Kingdom's Athena Swan program launched in 2005 to improve the representation of women in science, technology, engineering, mathematics, and medicine.

The Dimensions program aims to address systemic barriers not only for women but importantly for Indigenous peoples, persons with disabilities, racialized groups, and members of 2SLGBTQI+ communities, too. As institutions signed on and pledged commitment, I explained to faculty, staff, and students that I had left a gift with every one of their institutions: the Dimensions Charter. It was now up to the people to hold their institutions accountable for real progress on equity, diversity, inclusion, and belonging.

Because of numerous, brave individuals who kept raising the difficult, thorny, and often taboo issues of EDI in science over decades, a new conversation is now happening in Canada. I am no longer forced to make the case everywhere I go for why inclusion matters. It is becoming accepted that a diversity of perspectives generates great science; improves creativity, problem-solving abilities, and innovation; and improves the accuracy of clinical trials, among other advantages. And I no longer experience the uncomfortable silence or, worse, the negativity and hate that was common just a few years ago. One university official barked, "She may be the f***ing minister of science, but she's still a woman. You would have thought they could have at least gotten a man to be science minister. No man would ram this gender sh*t down our throats." Regardless, new offices and new positions for EDI are being created in institutions across Canada and in many other parts of the world. Numbers are changing in the Canada Research Chairs, and across research programs, conversations are taking place in institutions from coast to coast to coast, and the academic community is digging into inclusion.

Going forward, I want to see institutions act on the big-ticket items: tenure, full professorships, equal pay, and the end of abuse, discrimination, and harassment, which I will tackle in the next chapter. While important work remains to be done, I am already hearing from some corners that "EDI has a shelf life" and that "the pendulum will swing back and things will get back to normal." We must all ask ourselves what normal is. Do we really want to go backward, and when that push-back comes, how do we respond to ensure continued progress?

CHAPTER 6

Ending Abuse, Discrimination, and Sexual Violence

OVER THE DECADES of university education, three words summed up women's experience in the classrooms, labs, and the field: inequality, misogyny, and sexism. Those words also represent the roots of gender-based violence, as well as emotional, financial, physical, psychological, and sexual abuse and coercion, and the threats of such acts.[471]

Sixty-five years ago, sociologists Clifford Kirkpatrick and Eugene Kanin published a study about "Male sex aggression on a university campus." Over 55 percent of female respondents reported having been "offended at least once during the academic year at some level of erotic intimacy": 21 percent reported experiencing "forceful attempts at intercourse", and 6 percent reported "aggressive forceful attempts" with "threats or coercive infliction of physical pain". The authors speculated that men used secrecy and stigma to pressure and exploit women. They further suggested that "college girls" should be trained in "informed self-reliance."[472] While men pushed for access, it was women who needed the fixing.

Celebrated MIT molecular biologist and professor emerita Nancy Hopkins, who pioneered zebrafish to probe vertebrate development and cancer, recounts the story of working in James Watson's lab as a young woman, where Francis Crick felt entitled enough to walk in and put his hands on her breasts.[473] She recounted that she felt "very embarrassed, but for him, not for myself."[474]

Six biomedical research organizations—the Medical Research Council, Cancer Research UK, Wellcome, University College London, Imperial College London, and King's College London—came together and founded a flagship institute for discovery research in biomedicine. The name chosen was The Francis Crick Institute. Nobel Prize laureate Crick was a proponent of eugenics, which the Institute acknowledges: "He also expressed troubling views on eugenics, opinions that we do not in any way share."[475] I think that it is important that we bring all sides of history to light and make sure all voices are in the room when a name or honor is chosen. How many centers, departments, grants, institutes, prizes, spaces, etc., at universities are named after brilliant men, and what is the proportion of male honorees to that of women?

Sexual harassment was and continues to be rife in the sciences.[476] While Richard Feynman is recognized as a Nobel laureate, a "great scientist," and "a character," his predatory pursuits of women are often dismissed as "well known" or explained away as a product of his time.[477] He dated undergraduates, used prostitutes, slept with wives of colleagues, and considered women who refused to put out after he bought them drinks to be "worthless bitches."[478]

It's time to stop dismissing these behaviors as relics from a bygone era. To call any woman a "worthless bitch" is never and was never okay. It is also time to stop recognizing any male scientist's contributions as invaluable, and his wrongdoings less important. It can no longer be acceptable to put any such man on a pedestal. The glorification of conquests and secret laughter about them in quiet corners of the university only provides a social license for such behaviors to continue. Some professors still hit on undergrads, grad students, postdocs, and colleagues. Perpetrators can still find protection, as students fear retaliation, some colleagues turn a blind eye to abuse, and some administrators carefully and deliberately manage cases often with a view to making them go away.

In 1975, feminists gathered in Ithaca, New York, "to support a university administrator who had resigned because of a faculty member's persistent sexual advances."[479] It was during this incident the term *sexual harassment* was coined. At last, there was a name for the groping, kissing, touching, and other unwanted behaviors that led to women leaving their jobs or to being fired if they dared turn down their bosses. A movement to end sexual harassment quickly spread across the United States and Canada.[480] Distinguished university professor Constance Backhouse and writer Leah Cohen published the first book about sexual harassment in Canada: *The Secret Oppression: Sexual Harassment of Working Women*. The book featured seven case studies, including that of a graduate student whose supervisor suddenly "showed up at her house and kissed her without permission." The drive to eliminate sexual harassment rapidly gained legal backing. In 1980, "an Ontario human-rights tribunal ruled that sexual harassment was a violation of the Ontario Human Rights Code," and "in 1985, the Canadian federal government prohibited sexual harassment under the *Human Rights Act*."[481] In the United States, Title IX of the Educational Amendments of 1972 prohibited sex discrimination under any education program receiving federal financial assistance, and the Clery Act attempted to bring greater awareness to campus crime. There were other, subsequent changes in the laws and their interpretation.[482]

In the mid-1980s, public awareness of the crime of date rape became more widespread. In the United States, Mary Koss, professor of psychology at the University of Arizona, designed the first sexual experiences survey which was administered in 1984 and 1985 to more than 6,000 college students across more than thirty institutions.[483] Almost 8 percent of male students "volunteered anonymously that they had engaged in or attempted forced sex" and almost none of them considered it to be a crime.[484] Men had held women down against their will, without consent, and were convinced that it was not rape because they had faced no consequences.[485]

In the late 1980s, women pushed to change the sexist climate that existed at North American universities.[486] Zoologist Anne Innis Dagg published *Miseducation: Women and Canadian Universities.* She had been turned down to do fieldwork because she was a woman and she had been refused tenure and full-time professorships with the rationale that these were first to go to men who "had to support families."[487] Dagg described a McGill fraternity that "circulated a poster of a woman who appeared to be enjoying a gang rape" and "engineering students at a University of Toronto orientation event simulating the rape of a woman with a doll and a beer bottle."[488]

As mentioned earlier, the "Gordon House Nine" at Queen's University countered a "no means no" campaign against date rape in 1989.[489] Despite calls for the perpetrators to be punished, it was decided that it would be "unfair to punish a few students when a larger group was probably involved."[490] About a month after these events at Queen's University, fourteen women students were murdered at Montreal's École Polytechnique by a man. They were Geneviève Bergeron, Hélène Colgan, Nathalie Croteau, Barbara Daigneault, Anne-Marie Edward, Maud Haviernick, Barbara Klucznik-Widajewicz, Maryse Laganière, Maryse Leclair, Anne-Marie Lemay, Sonia Pelletier, Michèle Richard, Annie St-Arneault, and Annie Turcotte. They were brilliant, motivated students, and creative thinkers, and their lives were cut short by a gunman who blamed them for his failure to gain entrance into the engineering program and opened fire on them while screaming "You are all feminists."[491]

My memory of December 6, 1989, is the call from my mother to me in Scotland where I was a graduate student: "Sweetheart, they were your age. . . . Their mothers, I can't imagine! I'm sick, just sick. . . . Love, he killed them because they were women." My mom could not finish. My father picked up the line, his voice cracking: "We all want to hold our daughters a little tighter tonight; you're a long way off. Be safe, just be safe." Tears rolled down my cheeks as I explained, helplessly, to my university friends what had happened in Canada.

Sexual harassment pervades classes, field studies, labs, observatories, offices, teaching hospitals, and even polar stations. It is detrimental to people, impedes careers, and takes an emotional and economic toll.[492] Far too often, women are harassed out of science, with significant, negative impacts on the system.[493]

Despite the brutal aftermath of the Montreal massacre and annual vigils in honor of the victims of December 6, it was only in 2018, nearly three decades later, that the federal government committed to develop a national framework to address gender-based violence, a framework that was recently endorsed by the provinces and territories.[494]

In the United States, the Campus Sexual Assault Victim's Bill of Rights and the Violence Against Women Act were signed in 1992 and 1994, respectively. In 1997, the Department of Education's Office for Civil Rights (OCR) declared that schools must have sexual harassment grievance procedures and in 2001 the OCR "implored schools to stop harassment." In 2011, the Dear Colleague Letter, one of the primary communication tools, was issued to explicitly address that "Title IX requirements related to sexual harassment must also address sexual violence." In 2017, under the Trump Administration, the OCR reversed course and rescinded the 2011 guidance.[495]

In 2017, news broke about infamous Hollywood producer Harvey Weinstein.[496] Countless people from all walks of life began to come forward with stories of sexual abuse and various high-profile men were accused of sexual harassment that, in some cases, spanned decades. Tarana Burke had founded the #MeToo movement years earlier, but it now ignited a worldwide conversation about sexual violence across genders and races, toppling scores of powerful men who had until then been protected by various elements of the patriarchy.[497] Although individual and collective attitudes changed quickly, universities and colleges have been far too slow to respond, particularly in science.[498] Reports by the National Academies of Sciences, Engineering, and Medicine (2018) and one commissioned by the National Science Foundation (2022) confirm what women scientists through the decades already knew: science has a systemic sexual harassment problem.

In 2015 and 2019, the Association of American Universities (AAU) undertook a survey to examine the prevalence of sexual assault and misconduct at colleges and universities. More than thirty institutions and over 180,000 students responded to the 2019

AAU survey. Over 40 percent of students who responded reported experiencing "at least one sexually harassing behavior" since enrollment and almost 20 percent reported that such behavior interfered with or limited their academic experience. Among graduate and professional women who were sexually harassed, 24 percent of incidents were by a faculty member or instructor (compared to reports of 6 percent from undergraduate women). Similarly, for graduate and professional men, 18 percent were harassed by a faculty member or instructor compared to 4 percent of undergraduate men.[499] A survey conducted by the University of Texas found that 20 percent of undergraduate and graduate women science students reported experiencing sexual harassment from faculty and staff. More than 25 percent of female engineering students and more than 40 percent of all medical students reported experiencing sexual harassment. The Pennsylvania State University conducted a similar survey and found comparable results with approximately 35 percent of undergraduates, 45 percent of graduate students, and 50 percent of medical students reporting having experienced sexual harassment from faculty or staff.[500]

In Canada, over 65 percent of doctoral students and postdoctoral fellows who experienced harassment identified the perpetrator as "someone who had direct authority over them or who held a senior position in the academic community."[501] While disgraceful, this data is not surprising considering the academy is known to be second only to the military when it comes to reported incident rates of sexual harassment.[502] I think back to the young woman who worked in my ministerial office and was asked by her professor to twirl in front of him to get her mark in graduate school. The university did nothing until, several years later, it "needed proof" of his bad behavior. The professor eventually went on to serve in a higher position where he had even more power and more access to young women in one of the highest offices in the land.

Again, this systemic impact of male–female power differentials in the academy is now well-established and beyond dispute. I have heard many stories over the years of how the differentials have impacted the lives of women scientists. Graduate students depend on professors for their grades, field, lab, research, and teaching experiences, as well as for postdoc letters of reference and jobs. In other words, students depend on professors for their next steps in life and for their futures. The sexual predator tells the graduate student that she is nothing without him and that he can make her career. He threatens that if she dares to expose him, he will ruin her and her future. If that does not work, then there is always this residual threat: "Who do you think they'll believe, a respected, tenured professor or a graduate student?" He has done this before and the message to her is clear: she is expendable and he is someone special who will be protected. An expert. A leading scientist. A tenured professor.

Over the past decades, how many men have been protected by the mechanism of tenure, at the expense of students, graduate students—our children? How many times has someone in a university administration looked the other way, even if reluctantly or uncomfortably, because Professor X is a "great scientist," we need him, the university needs him, and "let's not forget he is doing great things for the world." No. A great scientist, any scientist, cannot prey upon young people. Anyone who does needs to be identified, investigated, and criminally charged if the evidence supports it. If found guilty, he needs to be removed from his tenured position and barred from teaching young people.

How many professors have yet to be called to account by the #MeToo movement, charged, and removed from the various academies? Will we ever know?

We cannot forget disgraced Michigan State University and former USA Gymnastics team doctor Larry Nassar, who was given an effective life sentence after being found guilty of abusing more than 150 gymnasts over two decades.[503] These women went to Nassar for help, but instead were sexually assaulted. When the women spoke up, they were ignored by three enormously powerful organizations, Michigan State University, USA Gymnastics, and the U.S. Olympic Committee.[504]

Dozens of women have come forward with legal claims against the FBI, alleging that the Agency knew by July, 2015 of Nassar's sexual abuse and assault, but failed to take action to stop him and protect athletes. Attorneys for the victims state that, "Nassar continued his predatory behavior, sexually assaulting approximately 90 young women and children between July 28, 2015, and September 12, 2016."[505] The Nassar case should have led to not only a global conversation on athlete health, safety, and well-being but also a discussion on campus safety. How dare any institution protect itself instead of protecting young people? How could the system so abjectly fail them over two terrible decades? It was only when his many victims, the women banded together that he was brought to justice. We owe them a debt of gratitude because they helped us to understand that this is not just a sport story.

Nassar's is one of the most notorious and well-known sexual assault cases, but it is only one case. We cannot pretend that it hasn't happened elsewhere. That it occurred on a university campus only makes it worse. One of the most important duties of an educational institution is to protect young people by providing a safe learning, research and sport environment. Every institution should remember Nassar's victims, and its leaders and managers should continuously ask themselves what more they can do.

It is not enough to hope we never see another Nassar again; we must make sure that the necessary safeguards are in place. The basics are to recognize that anyone can be a potential abuser; listen to victims; stop protecting colleagues and institutions; and follow-up on reported incidents. If a postsecondary institution fails to pick up on such extreme, horrifying behavior, what are the chances another would act on equally systemic, but fewer numbers of victims of abuse?

We must also remember that the perpetrator is not always a man. Women, non-binary, gender-diverse, and sexually diverse people can and do harass others and can be abusive. But in the context of women in science, it is often about men. In 2019 in the United States, 13 percent of students reported "non-consensual sexual contact by force or inability to consent;" the estimate for women undergraduates was nearly three times higher than for graduate and professional students (26 percent vs. 10 percent), and twice as high as the estimate for male graduate and professional students (7 percent vs. 3 percent). Over 7 percent of undergraduate women experienced "penetration by physical force" and over 5 percent by inability to consent. Over 90 percent of women and transgender, genderqueer, non-binary, or otherwise gender non-conforming students (TGQN) reported "at least one type of behavioral or emotional consequence" because of incidents with penetration.

Moreover, less than a third of women who reported non-consensual penetration by physical force or inability to consent made contact with a program or resource (30 percent), compared to almost 20 percent of men and over 40 percent of TGQN students. There are several reasons why women did not reach out for help. In some cases, they thought that they could "handle it themselves" or that the incident was not sufficiently serious to warrant help. In other instances, they felt ashamed or embarrassed or thought that it would be "too emotionally difficult" to seek assistance.[506]

The percentages are disturbingly similar in Canada: "45 percent of those who identified as women and 32 percent of those who identified as men personally experienced at least one unwanted sexualized behavior in the context of their postsecondary studies." One in ten women students experienced sexual assault in a postsecondary setting during the previous year and one in five of those said that the "assault took the form of a sexual activity to which she did not consent." These statistics are staggering. Eighty percent of women and 85 percent of men reported perpetrators were fellow students. "Relatively few" students said that the perpetrators were professors and others in positions of authority. Only 8 percent of women and 6 percent of men who experienced sexual assault spoke about the assault to someone associated with the school. Many did not think what happened was serious enough to report, did not know what to do, or did not trust how discretely or effectively the institution would handle the situation.[507]

Ask a woman faculty member if she has experienced sexual harassment over the course of her career and then ask her if she reported it, whether the complaint was taken seriously, and if any action was taken at all.

Here is my experience. As a university professor, I reported to the head of my department that a student had looked up my skirt. The chair refused to do anything about it. He made the incident my fault because of the way I chose to dress and excused the voyeur's deeply offensive behavior. I was violated.

When I started teaching in a tenure-track position, I was twenty-four years old and one of only two women in my department. I didn't have much in common with those around me, but I had a chair who cared, who was accessible, and who was kind to me—a real mentor. I quickly learned that there was another faculty member I should avoid. Every time I walked past his office to reach mine, he would walk out to stop me. The first few times, I thought it was nice; he was just trying to be welcoming. Then I started to note that he never missed an opportunity to interrupt me or to hold me back and that he stood too close. I had to keep backing up during any conversation. I started to feel accosted by him. I eventually realized that even if his door were closed, he would magically open it as I approached and stride into the hallway to detain me.

One day, I tried to make light of the situation, saying, "Your door was closed; how could you possibly have known it was me?" He replied, "I know your walk; I listen for your sexy walk. I recognize the sound of your high-heeled shoes." I had a sickening feeling. He was getting bolder. He was now looking at my chest and legs. I reminded him that, "my face is up here" to which he responded, "of course I know where your beautiful face is." I complained to my husband and my mom. I said, "No one will believe me, there's never a witness, and who will even care? He walks a fine line, and how can I prove that he's looking at my legs? I can't."

"And let's not forget," my husband said, "he has tenure and you don't." I learned to tip-toe past his office. I still got caught. "I think you're trying to sneak by me," he admonished. I changed to wearing running shoes. The final straw came when he actually touched my chest. I couldn't believe that he had done it. My shocked face must have said it all. He immediately looked scared and quickly uttered that "there was a hair on your sweater." Not damn good enough. And if the hair on my sweater was so offensive, tell me, don't touch me. I immediately went to complain. I explained what had been taking place. And while there was careful listening, I was told, "I've known him for many years. He's a good professor and a good person. He has a wife and children. Don't you think you mistook the situation?" I did not. And having a wife and children has never stopped anyone or anything. I once had to arrange counseling for a former student because her graduate supervisor, who had a wife and children, had engaged in an inappropriate relationship with her.

I explained that the faculty member "makes me feel uncomfortable and if he can do that to me as a colleague, how does he make students feel?" There was no resolution. On the way out, however, the assistant, in the adjoining room, called me over and shut both doors. "You're not wrong. You did not mistake the situation. This has been going on for years." She made a face and said, "He's gross. But you have to be careful; he has tenure, you don't, and you're a woman. If you cause a problem, you won't get tenure. I know how it works here. Keep your head down, do your work, and stay away from him."

I learned to walk another way to my office and I kept my door closed. It got better for a while until a new young woman joined the department and then it started again, this time with her. She came to my office, sobbing. She didn't know where to go for help. I went back, complained, and received the same advice: he's a good person; she's young and clearly not reading the signals right. I told her to keep her door closed, like mine, and that I would always be there for her, whenever she needed me.

And I was. I only wished I could have done more. And then a call came from the administration of the university. I can't remember the exact title of the person, but it was a position like a gender ombudsperson. I didn't even know we had such a position or what she did.

She explained that there was a problem and that they needed me to take a new thesis student. I explained that I already had a handful, that I met with them once a week for an hour each, and that I couldn't take another. The person said that this was a sensitive situation and that they really needed my help. I asked about the problem, but she refused to tell me. I then asked about the student, and she gave me her name. I had already pulled the same student into my office. She had lost so much weight that I was concerned about her and wanted to know if I could help in any way. When I asked the student what was wrong, she cried. She couldn't bring herself to tell me. My heart broke for her; she was so unhappy. I demanded to know the issue from the woman on the phone. If they wanted my help, they needed to tell me. I would not be complicit in some untoward solution. It involved that same male faculty member about whom I had complained twice. He had been the student's thesis advisor, and he had touched her—grossly inappropriately—in the secrecy of his office.

I felt the room spin. I had warned them. I had said that if he were bold enough to touch me, what would he do to a powerless student? I was furious and I told the woman on the phone everything. I asked whether she knew anything about him and whether any incidents had ever been reported. She wouldn't tell me. I demanded to know what the university was going to do for the student. "She won't press charges; there's nothing that can be done." I said, "Don't put this on the student. She'll be scared to death, she's about to graduate, and she'll just want to get the hell out." I asked again what the university would do and the administrator responded with the common refrain that the professor "had tenure." And there they were—those terrible words that protected a predator and protected his job for life. Who was there to protect the student? The professor serves students. The professor is there to provide, first and foremost, a safe and healthy learning environment and provide education, expertise, and opportunities. Students should not be sexual opportunities for professors.

On another occasion, when I was looking to leave the university, an older man asked me to come up to his executive office in one of our large, chartered banks. He told me that he was impressed by me, my "sharp mind," and my "ability to generate ideas." When I went to discuss the so-called job opportunity, unbeknownst to me, he locked the door behind me. When we sat down, he touched my knee. I removed his hand. He touched my knee again and I removed his hand a second time. I stood and he pressed himself against me. I pushed him away, hurriedly talked about his wife, and pointed to her picture. He didn't care. He pressed in again. I started to look for the exit, but he was between me and the door. I quickly moved around the desk and he followed. I was surprised by his agility and strength. He chased me around his office. To this day, I am not sure how I managed to escape. It remains a blur.

I ran to the elevator, I ran to my car, and it wasn't until I was almost home that I pulled over and vomited. I told my family as soon as I got home. They asked me what I was going to do. It was pointless. I had already had more than enough experience from the university to know that nothing would happen. I had complained before and I was told that I had misread the situation. Ironically, the man was linked to a university. He was rich and privileged, and

I am sure that this wasn't his first time. In contrast, I was a young academic and just starting out.

Hindsight is always 20/20. I played over and over in my mind what had happened in his office. The one thing that stood out in my memory was an older woman, an assistant who shook her head as I went into his office. I often wonder whether her action was meant as a warning or as a reproach, and if it was indeed a caution and she knew, why did she do nothing more? Perhaps it was because of her job; perhaps it was because she had already tried and nothing was done; or perhaps she, too, had learned that it was pointless? It's easy to look back, lay blame, and ask why nothing was done. But twenty years, ten years, and even two years ago were very different climates. Things are at last changing. National and international conversations are now taking place. Students, faculty, the university community, and parents are demanding that more must be done.

According to a 2018 National Academies of Sciences, Engineering, and Medicine (NASEM) report, sexual harassment can take three forms: gender harassment; unwanted sexual attention; and sexual coercion.[508] Gender harassment refers to "non-verbal and verbal behaviors that convey exclusion, hostility, objectification, or second-class status of one gender" and is by far the most common form of harassment. Gender harassment is not always recognized as a form of sexual harassment, although it can be equally damaging, and where it is pervasive other types of sexual harassment are also more likely. *Unwanted sexual attention* refers to "unwelcome verbal or physical sexual advances." *Sexual coercion* refers to cases in which favorable educational and professional "treatment is conditional on sexual activity" with the perpetrator.

According to the NASEM report, almost 60 percent of women faculty and staff across academic disciplines experienced sexual harassment, and women of color experience more sexual and racial harassment than any other groups. The report shows that when women are harassed, many ignore or try to appease the harasser and may seek social support. Many women do not report it because they correctly judge that they may experience retaliation or other negative outcomes if they do. A 2021 Statistics Canada report found that 34 percent of women and 22 percent of men employed at postsecondary institutions reported experiencing some form of harassment in the preceding twelve months.[509]

Sexual harassment undermines women's educational and professional achievement, impacts mental and physical health, affects job performance and satisfaction, and may cause women to leave the field.[510]

It is a shameful waste of talent. It is even more astonishing and disturbing that there is no evidence from the surveys that "any of the current policies, procedures, and approaches" have actually reduced sexual harassment. A focus on "symbolic compliance with the law" and "avoiding liability" has not worked.[511] The way we treat cases of sexual misconduct on university campuses remains an acute and systemic problem. Even people who are found guilty can often maintain their positions, keep their funding, get promoted, and train students, and the health, safety, and well-being of students, faculty, and the university community is undermined. Urgent changes are needed. We will not get to where we need to without institutions providing annual sexual harassment data to government bodies, focusing on prevention, and holding perpetrators to account—criminally when it is supported by the evidence. And all educational institutions must have a duty of care to students under which they, in turn, can be held to account.

A 2014 study of anthropologists and other field scientists found that close to 65 percent of more than 650 respondents had experienced some sort of sexual harassment while doing fieldwork.

More than 20 percent of respondents reported that they had personally experienced sexual advances, sexual contact, or sexual assault. About 20 percent of that group felt that it would have been "unsafe to fight back or to withhold consent" when they were sexually assaulted. Most victims were young, including students and postdocs, and fewer than 40 percent recalled a posted code of conduct at the field site. Only about 20 percent responded that they were satisfied with the outcome of their reporting.[512] While I am deeply saddened by these statistics, I am not surprised. I know firsthand what it is like to have to hide behind a door so that a colleague or someone in power does not keep staring at my chest or legs, and I know what it is like to be frightened for my safety out in the field. In one instance, while I was going about my work with a colleague, the work situation suddenly changed, and it became dangerous and menacing. We were in the middle of nowhere, there was no one else around, and he locked the car doors.

According to a 2017 study of almost 500 astronomers and planetary scientists between 2011 and 2015, women of color reported feeling unsafe in the workplace because of their gender or sex 40 percent of the time and because of their race almost 30 percent of the time. Moreover, nearly 20 percent of women of color and over 10 percent of white women skipped professional events because they did not feel safe attending.[513] I understand this, too. I stopped going to conferences because I was tired of the leering and hearing lines like this: "Oh, you're new. I liked your presentation, why don't you come up to my room later and we can discuss it?" Trying to be professional, I'd respond, "Because

I don't discuss research in men's rooms. If you want to talk about the results, I suggest we do it here in the open, like you would do with any of your other colleagues." The response: "Oh come on, darling, we all know why we go to conferences. It's to have a bit of fun on the side."

A 2022 study revealed that participants in the U.S. Antarctic Research Program suffer systematic and high levels of sexual harassment and assault: over 70 percent of women reported sexual harassment as a problem in the community, compared to almost 50 percent of men and 40 percent among leadership. Moreover, those working in the remote environment did not trust their employers to take harassment complaints seriously or to protect victims and punish perpetrators.[514]

There are running themes throughout the data: despite all the awareness over the last half-century, all the laws and programs put in place, the number of women impacted is overwhelming, and they remain uncomfortable about coming forward, fearing retaliation, and expecting that nothing will be done to address the harm they experienced. Until the employing universities and research centers take a zero-tolerance stand and until proven perpetrators are held to account—defunded, demoted, and dismissed, despite tenure—we will continue to lose women from science. Tenure can no longer serve as an excuse for the inexcusable. I asked a friend and former university administrator how many tenured professors had been let go during his time at a very large, well-known university. He thought long and hard about it and said, "I can think of only three over a period of about forty years. And one only because it had become a PR nightmare for the university."

Each university and college should feel compelled to take a hard look at itself and at its record. How many sexual harassment cases and sexual violence cases has it had? How many tenured professors have ever actually been removed from their positions and their universities for cause? And are there enablers in place that protect the organization and brand over people? What does tenure mean to a student, graduate student, or colleague? What is the sexual harassment and assault policy? Is there mandatory training? How often is the policy reviewed? And whom does it protect? Is the complaint taken seriously or is a case managed or, worst, swept under the rug? Are complaints tracked, repeat offenders noted, and do universities have to disclose and to whom?

As minister of science, I tasked our department with undertaking the first study of "harassment across postsecondary institutions." Under the title "Harassment and discrimination among faculty and researchers in Canada's postsecondary institutions," the results were published in 2021. Unsurprisingly, they revealed that harassment of both women and men is higher than that reported "in most other occupational settings." One-third of women and one-fifth of men experienced harassment in the previous twelve months, and almost 10 percent of women respondents reported unwanted sexual attention or sexual harassment in the year prior to the survey—"3.5 times higher than the proportion for male respondents." Almost half of respondents with disabilities experienced some form of harassment in the year prior, although self-reporting likely underestimates prevalence. And while First Nations, Métis, and Inuit accounted for 2 percent of the academic community respondents, "40 percent experienced humiliating behavior, verbal abuse, or another form of harassment" in the year preceding the survey.[515] The collection of data alone won't trigger change. Simply telling a story won't lead to change. This information must be used to improve lives—the next step will be to develop new and effective measures to protect all women who have chosen science as their path.

In Canada, sexual violence at post-secondary institutions is increasingly recognized as a serious problem. In 2015, the Council of Nova Scotia University Presidents put forward a Memorandum of Understanding with Nova Scotia Universities to improve measures to prevent sexual violence. In 2016, Ontario enacted Canada's first legislation: Bill 132, the Sexual Violence and Harassment Action Plan Act. British Columbia followed with legislation in 2016, Quebec in 2017, and Manitoba and Prince Edward Island (PEI) in 2018. Manitoba became the first province to offer centralized online reporting of sexualized violence at post-secondary institutions, and PEI became the first to limit the use of non-disclosure agreements. In 2020, the New Brunswick government convened a roundtable concerning sexual violence on campus and released the findings in 2021. In 2022, a Ministerial Letter was issued to all public, post-secondary institutions in Alberta to review and revise their policies and procedures.[516]

We have seen the numbers of faculty, students, and the post-secondary community impacted. Every delay to act matters, and everyone should have access to the same minimum standards of justice.

Governments must send a clear signal that they not only care about sexual harassment and sexual violence on campus, and in university laboratories and residences, but also demand real improvement. That means that they will need to know baseline data: what is the prevalence of sexual harassment and sexual violence, who is being affected, and where is it occurring? Clear terminology and "aggregate data are key." These encourage consistency, accountability, and transparency among institutions and protect individual case data and privacy. In Canada, real questions abound regarding institutional monitoring of policies, data collection, processes to address sexual harassment and sexual violence, and enforcement of their policies because institutions often have prohibitions about sharing information internally and externally. These prohibitions continue to stand in the way of effective solutions.[517] Over the years, there has been systemic resistance to such data being submitted as institutions work to protect their brand and competitive advantage. Surely, in 2024, student, faculty, and community health, safety, and well-being will trump the need for image.

I would like to see a requirement that institutions submit annual data on sexual harassment and violence case numbers and how these cases are being addressed to governments. Additionally, assessing whether people feel the processes in place are effective, among other indicators, should be included. Funding agencies must also send a signal to the academic community that they are serious about tackling sexual harassment, as the National Science Foundation has done, "and academic institutions should consider sexual harassment equally important as research misconduct in terms of its effects on the integrity of research."[518]

Governments could go further by offering institutions a model sexual harassment policy that protects the rights of survivors while giving alleged harassers due process rather than immunity. NASEM (2018) recommended that institutions have clearly stated policies with disciplinary consequences for perpetrators, and that punitive actions correspond to the severity and frequency of the harassment. Moreover, any investigative process should be fair, and decisions should be taken in a just and timely way. Institutions need to publicly commit to a zero-tolerance climate for sexual harassment.

"The extent to which an organization's climate is seen as permissive of sexual harassment stands in the strongest causal relationship to how much harassment occurs in the organization."[519]

If survivors choose to break the silence and speak up, they let others know that they are not alone. They empower others to do the same. The courage of a victim speaking out is also a public expression of her trust that she will be protected, that accountability structures are in place, and that justice will be achieved. It is up to universities to step up and do their part to honor and reinforce that trust.

Academic institutions must accept and value the courage behind a report of sexual harassment, and they must understand that formal reporting can sometimes revictimize people. It is therefore important to develop alternative reporting processes. Institutions should also provide victims with healthcare, legal services, and support services and take measures to prevent retaliation.[520]

Academic leaders need to pay increased attention to gender harassment, to address the most common form of sexual harassment and to help prevent all other types.[521] That means zero tolerance on comments like "young lady," "the only reason you got this position is because you are a woman," "arguing with you is like arguing with my wife and mother," or any other lewd comments from colleagues or students.

When women students or faculty complain about a colleague's or a student's behavior, there needs to be a follow-up rather than a dismissal of her concerns. It was never okay for a student to look up my skirt and have it dismissed by the department head as "Well, you do wear short skirts" as he walked out the door. What I was wearing should have had no bearing whatsoever. That's called victim-blaming. Academic institutions need to build "diverse, inclusive, and respectful environments" and that means hiring and promoting more women and women from diverse backgrounds. Rewarding "collegiality and cooperation" alongside individual research and teaching "could have a significant, positive impact on institutional climate," particularly as some colleagues confuse academic rigor with cruelty.

Universities should dismantle the "hierarchical," often financially driven, relationship of power asymmetry and dependency between faculty and undergraduate and graduate students and postdocs.[522] Committee-based advising or mentoring networks have been identified as productive mechanisms.[523] In an academic setting, supervisors and trainees sometimes work alone, late at night in an empty lab, go for drinks afterward, or may have side-by-side tents at a field camp in the Arctic. Sport, equally rife with asymmetrical power structures, offers some protection through the Responsible Coaching Movement and the rule of two: the requirement to have two certified coaches present in any potentially vulnerable situations such as closed-door meetings, travel, and remote training environments, particularly with underage athletes. It means that someone else is always hearing and watching, but it only works if that second person is fully empowered to speak out, intervene, and report inappropriate behaviors or comments.

To eliminate sexual harassment, abuse, and assault in the academy, government, institutions, academic leadership, faculty, students, and the university community must work together. After all, conversations about sexual harassment have been happening on campuses for the past fifty years. It's not enough to simply check a box in a survey or an annual review. We need real, ongoing action, daily efforts to maintain safe spaces, enforcement of policies, and apologies, and we need the numbers to change. That means each of us must commit to a duty of care to students and colleagues, and all of us must refuse to enable those who harass us. It means calling out bad behavior for the sake of everyone in the academic community, identifying and isolating the abusers, and consistently enforcing consequences for all sexual predators and other perpetrators, even if they happen to be "great scientists." Nothing less will do, and when we get there, we will have a safer, more inclusive, and stronger system—one that can focus on its core aspiration: scientific discovery for all of humankind.

CHAPTER 7

Leading Change

THROUGH TIME, MEN WERE the chosen sex. Their leadership and participation in society were considered central to the natural order, attributable to their innate attributes. Men came up with a myriad of reasons why women were lesser. They had smaller brains. They were emotional. They had wandering uteruses. They were hormonal.[524] According to men, women simply weren't fit to contribute in any meaningful way, and they were certainly not built to lead.

Women were kept from positions of leadership in the academy as long as possible and when it was no longer possible to exclude them, they were treated abysmally. In 1884, Letitia Salter was hired as the first "Lady Superintendent" at the University of Toronto and placed in charge of nearly 200 students. Seen as an unnecessary strain on the university's resources, she was paid less than the university president paid his coachman.[525]

In 1974, Pauline Jewett became the first woman to head a major Canadian coeducational university, cracking the glass ceiling of academic leadership. After serving as a professor of political science and member of parliament in Canada's House of Commons, she served as president of Simon Fraser University until 1978.[526] In 1974, biochemist Lorene Lane Rogers is believed to have become the first woman to lead a major American university. Early in her career at the University of Texas, a member of the chemistry department told Rogers, "We've never had a woman in our department, and we never will." Undaunted, Rogers worked her way up the academic ladder until she was the institution's vice president. When she was appointed as interim president of the University of Texas in 1974, faculty and students protested in an effort to force Rogers to resign. She held on and led the University of Texas between 1974 and 1979.[527]

Instead of the dissent and demonstration that accompanied the rise of Lorene Lane Rogers, I wish that there had been more of what Father Shook noted regarding women at St. Michael's College, a federated college within the University of Toronto: "The fact remains that every time a major concession was made towards the fuller integration of women into the college, a distinct advance in academic excellence followed."[528]

From the mid-1970s, it would be another two decades before a woman would lead an Ivy League school. Judith Rodin became the first woman, permanent president of the University of Pennsylvania in 1994. At the time, 12 percent of the 3,200 American college presidents were women and just 9 percent were headed by minorities.[529]

After another two decades, Louise Richardson became the first-ever woman vice-chancellor of Oxford University in 2016, ending a succession of 271 male vice-chancellors and nearly 800 years of male leadership.[530] In 2022, Queen Elizabeth ll appointed Richardson a Dame Commander of the Most Excellent Order of the British Empire (DBE) in recognition of her service to higher education. I have always wanted to believe that once the university and the broader society witnessed a strong powerful woman accomplish what was until then impossible, the walls would come down for others. And encouragingly in 2023, "one of the oldest and most prestigious universities in the United States" Dartmouth College, announced that it would welcome Sian Leah Beilock as its first woman president after eighteen presidents and more than 250 years.[531] In the same year, Harvard University announced that Claudine Gay would become its first Black president and second woman in the role.[532] According to an analysis of world university rankings by *The Times Higher Education*, the share of the world's top universities that are led by women reached 20 percent for the first time in 2021.[533]

While these appointments fill me with hope, I know that women must continue to prove their worth and that their leadership paths remain more difficult and potholed than those of men. Moreover, men continue to over-estimate women's representation in top positions and threaten to halt progress at current, unacceptable levels. The proportion of women presidents was just 14 percent in 2016.[534] If that rate of increase continues, it will take decades to reach parity. And this assumes that social progress is linear rather than something that comes in fits and bursts, and often with real push-back—the hallmark of integration of women and minorities into higher education. The bottom line is this: we need more women entering the leadership pipeline and the pace of sustained change must improve.

The percentage of women college presidents has increased over the last thirty years in the United States, from roughly 10 percent in 1986 to a plateau of 25 percent to 30 percent in the last decade—although initial growth reached 19 percent in 1998.[535]

Women typically follow different paths to the presidency than men, with 32 percent of women presidents having altered their career progression to care for a dependent, parent, partner, or spouse. Over the last three decades, the percentage of minority college presidents has slowly increased, but "women of color" continue to be the most underrepresented in the presidency at just 5 percent.[536]

Only 20 percent of the world's top 200 universities have female presidents and it took over 760 years for Oxford University to hire a woman vice-chancellor: Louise Richardson.[537] Women held about 30 percent of university president posts in Canada, but this number dropped sharply to only 13 percent when it came to the presidents of the research-intensive universities. Since 2010, 61 percent of unfinished presidential terms were held by women. Something is clearly wrong here. Women are not only underrepresented but also disproportionately over-represented when it comes to shortened mandates. This trend is concerning and supports a "long-standing belief that the risks associated with university leadership are more significant for women."[538]

Martha Piper, the former president and vice-chancellor of the University of British Columbia, jointly authored a book with Indira Samarasekera, the former president and vice-chancellor of the University of Alberta. In *Nerve: Lessons in Leadership from Two Women Who Went First*, Piper reflects that "being the first is not good enough, that until we have the second, the third and fourth women leaders, one after the other, we will continue to view women in leadership as the exception rather than the rule."[539]

While universities proudly name each new woman president, they also brand and mark her as a "woman," not just a "president," as with men. This marketing cuts in several ways. Institutions are shown to be modern and progressive; women are celebrated and highlighted, inspiring this generation and future generations; however, women are also shown to be different—a woman leader and not just a leader. And in a world where women's leadership remains underrepresented and is still called into question, emphasizing her leadership can be seen as outside the norm, a nod to the left and the politically correct, and it may encourage the perception that women are taking over at the expense of men, although the numbers show the opposite.

The cards can be stacked against her before she ever enters her new office and her stakeholders may choose to interpret how she leads, how she makes decisions, and what she focuses on, according to her being a "woman." If she crashes, some may interpret "her" failure, as further evidence that women are not meant to lead. In decades past, it was common for women leaders to be hired amongst a homogeneous group of male leaders and given leadership roles that were precarious and doomed to fail.[540] Even today, few institutions are hiring a second woman in the role of president.[541] The question we should all be asking is whether women are getting the same quality of promotions as men.

Malinda Smith, today vice provost and associate vice president of research (equity, diversity, and inclusion), University of Calgary, looked at the diversity among the academic leadership of Canada's most research-intensive universities, the so-called U15. She found that leadership remains predominantly male and white, with "scant representation" of visible minorities or Indigenous peoples. U15 board chairs were almost 60 percent male and over 85 percent white; less than 10 percent were minority female and none were Indigenous. U15 chancellors were almost 75 percent male and 100 percent white. Moreover, these senior leaders were drawn largely from a narrow group of professional fields, primarily the corporate sector, particularly banking, finance, and corporate law.

U15 presidents were also over 85 percent male and 80 percent white, and U15 deans of faculties and schools were almost 70 percent male and over 90 percent white. Under 10 percent were visible minority or Indigenous. Presidents, provosts, and vice presidents (academic and research) were primarily selected from STEM disciplines, where there is a lack of women. Smith's work shows that "despite three decades of employee equity initiatives to advance the diversity of peoples and perspectives" and better reflect Canadian society, the leadership of Canada's most research-intensive universities remains male and pale.[542] Data from the United States is like that of Canada, with white males filling most leadership positions. In fact, more than two-thirds of college and university presidents have been men and over 85 percent of administrators are white. Only 7 percent are Black, 2 percent are Asian, and 3 percent are Latin.[543]

I find it interesting that the role of the university president is changing from that of an academic leader, traditionally chosen because he was thought to be the "smartest man in the room" and therefore born to run a large, complex multistakeholder organization, toward a more CEO-type in recent years. The president is responsible for setting the vision, strategy, and long-term objectives of the institution and transforming it into the future. She, he, or they must also navigate—along with the senior leadership team—the unprecedented challenges, including governance and accountability, unpredictable political environments, lack of financial resources, international outreach, complicated stakeholder relations, shifting student expectations, to name just a few issues. All this is changing, but the attitudes are not.

Women leaders can provide example after example of where the credit, strategy, or execution of a plan went to a man for her work, how she was intentionally or unintentionally undermined, why she did not get that second term, and why a woman did not follow her. Throughout my career, I have longed to hear women academic leaders described in terms like intelligent, visionary, strategic, intentional, deliberate, accomplished, successful, and changemaker. But I have not. Rather, women were described as compassionate, enthusiastic, passionate, warm, and exotic—such was our Svalbard research team's favorite word for me, meaning I was different, an outlier, and not one of them. Repeatedly, I witnessed woman after woman offered coaching to improve her leadership skills and to take risks, while the institution failed to recognize that the workplace culture is unequal, women are treated differently than male colleagues, and men are often rewarded for risk-taking, while women are punished for the same behavior. Institutions are still failing.

* * *

Women were similarly excluded for as long as possible from participation and leadership in politics. It wasn't until the early twentieth century that women in Canada were recognized as "persons" by their own federal government. Alberta's "Famous Five"—Emily Murphy, Nellie McClung, Irene Parlby, Louise McKinney, and Henrietta Muir Edwards, who had all been active in women's rights dating back as far as the 1880s—initiated the ground-breaking Persons Case (Edwards v. A.G. of Canada). In 1927, with support across the country, they petitioned Canada's Supreme Court. After weeks of debate, the court ruled that women were not "persons" according to the British North America Act and therefore were ineligible for appointment to the Senate. The women refused to accept the ruling and appealed to the Privy Council of England, Canada's highest court at that time. In 1929, the Privy Council reversed the Supreme Court's decision and said: "No, we should recognize women as persons." [544]After millions of years of human evolution, millennia of oppression, servitude, and ignorance, some women in Canada were at last recognized as persons.[545] First Nations women, for example, did not receive the right to vote until 1960.[546] It is important to remember that some members of the Famous Five were also anti-immigrant, racist, and supportive of eugenics.[547]

While the Privy Council's extraordinary ruling was limited, the political winds were shifting. The win meant that some women could in fact be appointed to the Senate of Canada, and some women could no longer be denied rights based on a narrow interpretation of the law.[548] Still, women's rights challenges remained. Over half a century later, Canada's Cold War fallout shelters protected mostly men. In 1984, over 300 people participated in a nuclear dress rehearsal in Atlantic Canada's only government fallout shelter, the Debert bunker. The coveted spots were for high-ranking government, military, and media, the majority of whom were male. Women questioned how men would feel walking out on their wives and children during a nuclear attack and whether protecting the male-dominated government would in fact protect humanity. Some suggested rewriting the plan to include females of child-bearing age and a sperm bank.[549]

It should be abundantly clear to everyone why having women lead governments and represent communities in their parliaments matters. Women account for 50 percent of the population. Their agency, voices, and ideas can have a profound impact on their countries' policies. We can work to help make a better country and world. But as of January 2024, there are just twenty-six countries where women serve as heads of state and/or government.

Based on the current rate of change, and again, acknowledging that social progress is not linear, it is projected to take another 130 years for gender equality to be reached.

More damning, only thirteen countries have women in 50 percent or more of cabinet positions. That's less than the number of countries where women hold the highest position in the land. Globally, just over one-fifth of cabinet positions are held by women. Typically, the most commonly held portfolios by women cabinet ministers are "gender equality, family affairs, children's affairs, social inclusion and development, social protection and social security, and Indigenous and minority affairs."[550] The net effect is to identify the issues in these portfolios as women's issues, when in fact they concern every family, community, and economy in a country, while also keeping women out of financial, international affairs, and defense portfolios.

In Canada, it took 148 years to have a gender-balanced cabinet and 153 years to have a woman finance minister. Having women at the table sharing their lived experience and the experience of women in their communities and across the country, changes cabinet discussions, how governments spend money, and, indeed, the national conversation. From the start of the global COVID-19 pandemic, it was clear that the economic fall-out affected women more than men.[551] In general, women had higher job losses, lower wages, and increased responsibilities for caregiving and children's education—it was dubbed the "she-cession." Canada saw a window of opportunity to push for and deliver on a national early learning and childcare plan because its government understood that without childcare, parents, typically women, can't work. The people around the cabinet understood the importance of the initiative because their grandmothers, mothers, and many of them had lived through it, and they'd heard the stories of women struggling from across the country.

There can be no doubt today about the positive impact of women in leadership.[552] Women, from Malala Yousafzai to Greta Thunberg to Jane Goodall, are powerful forces for good in the world. Women's leadership has also been shown to positively impact the bottom line of business and to help organizations make more ethical decisions. Yet 2022 data from the Reykjavík Index for Leadership, "an annual survey that compares how men and women are viewed in terms of their suitability for positions of power," shows that only 45 percent of respondents in the G7 are "very comfortable with a woman at the head of their government, down from 52 percent in 2021," and 47 percent say they were "'very comfortable having a woman as CEO of a major company in their country, down from 54 percent a year earlier." Men are significantly more likely than women to be critical of a female leader.[553]

Slightly more than one-quarter of parliamentarians are women globally, up from just over 10 percent in 1995. "Only six countries have achieved 50 percent or more women in parliament: Rwanda (61 percent), Cuba (53 percent), Nicaragua (52 percent), Mexico (50 percent), New Zealand (50 percent), and the United Arab Emirates (50 percent)."[554] While Canada reached a record 31 percent of women elected in the 2021 election, Canada ranked fifty-ninth worldwide in terms of women's representation in national lower or single houses of parliament.[555] The United States reached 28 percent in the 118th Congress.[556] These figures should concern us all. What they show is that the centuries-old pattern of male leadership remains alive and well and that the fear of the zero-sum game—if women succeed, men lose—remains.

In 1962, Vice President Lyndon B. Johnson famously killed the Mercury 13 program by simply scribbling "Let's Stop This Now" on a memo to NASA.[557]

Mercury 13 had been a privately funded program with thirteen stellar women pilots. Its aim was to test their suitability for space travel. They included Jerrie Cobb, Bernice Steadman, Gene Nora Jessen, Irene Leverton, Janet and Marion Dietrich, Jane Hart, Jean Hixson, Jerri Truhill, Myrtle Cagle, Rhea Woltman, Sarah Ratley, and Wally Funk.[558] To give a sense of their accomplishments, Cobb, alone, had flown her first plane at age twelve, started teaching men to fly at nineteen, and began delivering military planes to Air Forces worldwide at twenty-one. By age twenty-nine, she had flown over sixty types of propeller aircraft, accumulated over 7,000 flying hours, and racked up flying records for altitude, distance, and speed.[559] The Mercury 13 women passed the same battery of seventy-five rigorous physical and psychological tests as the Mercury 7 male astronauts—NASA's original chosen seven in 1959. Candidates had to swallow long rubber tubes to test their stomach acid, endure nerve tests, and even spend time in an isolation tank. Cobb tested in the top 2 percent of all astronaut contenders.[560] Funk lasted ten hours and thirty-five minutes in the isolation tank without hallucinating, surpassing the male record of just over three hours.[561]

Women had clearly proved that they had the right stuff, but they were the wrong gender. When the program was abruptly canceled, Cobb and Hart, a mother of eight and the pilot and wife of Senator Philip Hart, flew to Washington to change hearts and minds. Hart testified before Congress: "It will perhaps come as no surprise to you that I strongly believe women should have a role in space research. In fact, it is inconceivable to me that the world of outer space should be restricted to men only, like some sort of stag club."[562]

Astronauts John Glenn and Scott Carpenter also testified but argued that women could not qualify as astronaut candidates; contenders had to be graduates of military jet test-pilot programs, and only men could do that.[563] While the women's case captured the imagination of the American public, the political winds were against them. Lyndon Johnson shuttered the program and a year later, in 1963, the Soviets launched cosmonaut Valentina Tereshkova into orbit. It was a space race defeat for the United States, a blow for American women, and it would take another twenty years before NASA sent Sally Ride on the space shuttle Challenger.[564]

* * *

To be fair, there have also been women who have sometimes stood in the way of progress. Since the 1970s, they have divided women and offered a stark choice between being a wife and mother or a feminist, raising fears of women being forced into the roles of combat soldiers and having to use unisex toilets.

Women are smart, intelligent, caring, and capable of deciding their own destinies. We do not need anyone, male or female, to curtail or limit our choices. As an early career researcher, I did not see the other woman in my department. I had no women mentors, no women in my network, and no women as colleagues to ask for career advice. I did my best to muddle through. As I progressed in my academic, business, and political career, I learned a tough lesson, one that we don't like to openly acknowledge or discuss: not all women want the best for one another. My experience is that when the numbers of successful women are few, competition can be fierce, with women buying into traditional power structures and the zero-sum game: if another woman gets ahead, I fall behind.

Throughout my career, I have witnessed women saying the right things, befriending young, smart, up-and-comers, and building trust with them, only to later build their power base by deliberately backstabbing, undermining, or publicly slamming them in front of colleagues. As women, we need to stand up for one another, build connections and relationships, amplify other women's voices, and call out bad behavior by people of any gender.

I have always wondered at what point we will see this disturbing and distasteful behavior disappear and reach a point where women want to not only succeed in their careers but also ensure the collective success of all women. During my time as minister, I had one bad experience following a conversation with two powerful stakeholder groups that had asked how they could help me after I had delivered a large tranche of funds for the research sector. I saw the opportunity to broach equity, diversity, and inclusion with both. Immediately, they were uncomfortable. The conversation became difficult and halted.

Eventually, one woman leader got on board and invited me to a special meeting with her impressive network. In good faith, I attended and started what I hoped would be an open discussion on inclusion in research. It did not go well. I felt like I was back in the 1990s. Silence followed my opening pitch. Backs stiffened around the table. Two other women in the room hung their heads and did not come to my defense.

Throughout my career, I have known some key woman leaders, most often from the STEM disciplines, where they have been inculcated in patriarchal culture, male power structures, and male-centerd labs. They not only accepted the status quo, but believe that it is up to women to make it on their own, just as they did. These same women often remove themselves from any discussions about women's representation and side with the majority male colleagues. Fortunately, one brave man spoke in my defense, and if it were not for his intervention, it would have been difficult to push universities to act on the long-standing EDI problems. One person had the guts to speak up when everyone else looked away.

I shared this story with a respected male leader in the research community. He said, "Well Kirsty, what do you expect? You're a woman; you look like a woman. It's the first thing people see as you walk toward them and then you go and talk about women's stuff. It doesn't belong in science." I patiently responded that when I sat around a table with a male like himself, "I saw a scientist, a colleague, someone with whom I could have important discussions about funding research and building a better research sector and Canada that included all people."

In the meeting with the stakeholder groups, I had expected women, especially, to stand up for one another, but not everyone is at the same place when it comes to inclusion. It takes time, work, and sustained discussion to bring along everyone, to say nothing of the people who deliberately obstruct. But if we want more women leaders, we need to come together, build allies, and support one another in good times and, especially, in bad. When people are fearful, whether because of a pandemic or a recession, they often look for security and stability and that has traditionally meant male leadership. Some media outlets will question whether women should be in leadership positions and view instances where women fail in leadership roles as evidence of a lack of drive, courage, or talent. Thus, we undermine a century of change.[565]

* * *

I was the first and only woman scientist to serve as Canada's minister of science. It became very important to me that every young woman knows that serving her community, taking her seat in the House of Commons, and leading change is a real possibility. Because I came from the world of research, I knew the players and issues in the scientific community, and I knew what needed addressing. I was disappointed when there was no mention of EDI in my ministerial mandate letter. I pushed back against the clear omission, but my plea fell on deaf ears, either because the prime minister's office did not know the issue or because there was fear about rocking the boat with stakeholders. I thought there was a clear point to be made that Liberals care about science, particularly after the Harper government had cut science funding and dismissed and muzzled scientists.

I appointed top-level women academics to key leadership positions so that there would continue to be strong voices at the table and continue to keep the pressure on. I highlighted the importance of EDI and gave universities clear signals that I expected their sector to change and work to build inclusive institutions. Not "inclusive excellence," as some stakeholders put forth. After all, it was just "excellence" when male scientists were the majority and no modifier was needed. And this imperative had to be translated to every university and college. It means getting the basics right if we want to see change in the top university role: job design and advertising; the role of external consultants; the diversity of selection committees; and the role and composition of the board. A lack of diversity among the board and selection committee disproportionately favors candidates who are white men.[566]

I can speak only from my own perspective, and I leave others with their lived experience to tell their important stories, but I have witnessed over decades the inexcusable treatment of women researchers from non-traditional backgrounds and have continually stood up for change as an academic, MP, and minister. While it is important to recognize that there have been real gains for some women in science, academia, and leadership positions, it has not been equal across the board and much work needs to be done going forward to ensure that everyone can reach the office of the president.

Today, throughout democratic societies, the importance of diversity is increasingly valued. Numerous research studies, for example, show that diverse teams increase customers, revenue, and profits in business. We see universities creating senior leadership EDI positions to track progress, improve culture, and create fair and equal environments for all. They are beginning to appreciate that EDI leads to increased innovation and also means more voices and better representation, processes, decisions, and outcomes. But universities have nevertheless lagged in this area, pushing the fear that diversity will dilute excellence and interfere with compliance with laws and regulations. Only recently have leading, peer-reviewed academic journals and institutions genuinely begun to embrace and value diversity.[567]

Let's be clear, the university remains a problematic workplace. While research and lived experience demonstrate that qualified and ambitious women are plentiful, women—especially racialized women and women from other backgrounds—are underrepresented in tenured and full professorships, which limits their opportunities for leadership positions. While some women are being hired as senior leaders, research and surveys of representation consistently overestimate women's representation. And if women do attain a leadership position, they often hold the least senior administrative positions and they are the lowest paid. It is past time to stop questioning and fixing women, their decisions, and their leadership. It's time to fix the university and maybe if we do this, we will see more women presidents finish their first term and even have a second term, with more women presidents following.

As we build this future together, we must all be wary of over-optimism as well as pushback from some dark recesses of the university. I am already hearing people say that EDI has a shelf life. That the pendulum will swing back and things will return to normal. Normal? What is normal? Do we really want to go backward? We have centuries of colonialism, slavery, and patriarchy to undo.

And yet in June 2023, the U.S. Supreme Court struck a severe blow against affirmative action. In response, President Biden wrote on X, "For decades, the Supreme Court recognized a college's freedom to decide how to build a diverse student body and provide opportunity. Today, the Court walked away from precedent, effectively ending affirmative action in higher education. I strongly disagree with this decision."

The Association of American Universities President Barbara Snyder also expressed regret: "I am deeply disappointed in today's decision by the Supreme Court to upend more than forty years of precedent by banning colleges and universities from considering race among other factors in their admissions decisions to create diverse student bodies and fulfill their academic missions." She added that the AAU would "continue its work to advance equity, promote inclusive environments, and speak out on the value of diversity on our campuses."[568]

On October 3, 2023, Michelle Donelan, the UK secretary of state for science, innovation, and technology, told the Conservative party conference that the Tories are "depoliticizing science."[569] The party's announcement explained that it was "kicking woke ideology out of science," thereby "safeguarding scientific research from the denial of biology and the steady creep of political correctness." Two days later, more than 2,000 researchers responded by signing an open letter expressing anger and concern that, "far from depoliticizing science, this policy appears to be driven by ideology."[570]

In November 2023, Alberta's United Conservative Party members voted unanimously at their convention to eradicate diversity and inclusion offices at universities and colleges.[571]

And in December, I received confirmation that the internationally recognized Dimensions program had ended. It was flatly impossible to fix EDI in a handful of years. Millennia of patriarchy, and centuries of colonialism and slavery cast long shadows. This is not the time to stop, but rather to continue, and fix well-known, long-standing inequality.

We cannot afford to go backward. When this push-back comes, we need to respond and ensure continual progress. That begins with remembering that progress doesn't happen in a straight line. Our culture evolves. Our understanding of issues changes through time. I have seen commitment to women's issues ebb and flow in my lifetime.

I was in elementary school in the 1970s when women's issues were front and center on the television. I remember the lead-up to the so-called "Battle of the Sexes" when tennis great Billie Jean King agreed to play Bobby Riggs, the former number-one ranked player in the world, then in his mid-fifties and a self-proclaimed male chauvinist. King understood that losing his challenge was not an option, prepared diligently, won in straight sets, and changed women's tennis and society.[572] I remember the first women's strike, the Equal Rights Amendment, *Time Magazine* awarding its annual "Man of the Year" to American women, the UN declaration of the International Women's Year (1975), and the Decade for Women (1976–1985). My family promised that life would be different for me. My grandmother said that I would go to university and wouldn't have to work multiple, terrible jobs and miss watching any children I might have grow up just to put food on the table, as she had been forced to do. My mom promised that I would be financially independent and wouldn't need a husband to assure a bank that I could possess a chequebook and be trusted with the money I earned. My father insisted that my brother and I were equal in every way and set the same expectations of both of us.

But I also remember the 1990s, when growing numbers of people did not believe we needed feminism and perpetuated inappropriate jokes and stereotypes against women. Progress on equality stalled considerably. Governments withdrew support for equality, the Canadian Advisory Council on the Status of Women was eliminated, and women's movements were on the defensive. The term "political correctness" began to be used by right-leaning groups to oppose what they perceived as the rise of a liberal agenda and the introduction of liberal curricula on university and college campuses. The changing political winds chilled campuses. Some conversations were pushed underground or became taboo, like pay equity. Because of continuing inaction, the gender pay gap for faculty in Canadian universities remains significant, with women professors earning on average 10 percent less than men for the same work.[573]

Progress on women's issues is also influenced by international and national agendas. These determine what actions governments, society, and institutions take, and that, in turn, influences how we think about one another. As a minister, I was able to hold the feet of academic institutions to the fire and put in place protections to safeguard against backward slippage over time. But ministers come and go, and each minister will implement not only their ministerial mandate but also what matters to them. Will future ministers put the same focus on equality for women in research or will this fall off the national agenda? Will future politicians? The progress I made with the Canada Research Chairs narrowly escaped undoing when a member of parliament forced a vote on reviewing the program's criteria for funding.

The point is political winds can change quickly. We must pay attention to politics, be vigilant, and remain committed to ensuring inclusion for all. We must also measure progress through data and metrics. Numbers matter. How many women land in senior leadership positions, and what is their average tenure in those positions? Do they finish their mandates or are they truncated early, and do they go on to serve in similar positions in other institutions?

It is up to institutions to collect, measure, and analyze data— after all, this is what researchers do. They must chart progress and decide what programs universities, colleges, and hospitals will put in place to address culture. What has changed in the institution and is there a new and properly staffed EDI office? What research is being led, and who is doing the research? Are women and diverse scholars yet again being tapped to do the heavy lifting or is the workload being shared by all? We must insist on at-pace, sustained progress, rather than the meager step-at-a-time advances we have all been encouraged and trained to accept and even celebrate. None of this is easy. Moving beyond past biases, old beliefs, and stereotypes requires us to tackle thorny cultural issues around inclusion and make the hard commitments that open leadership roles to everyone.

I recently had a conversation with an early career woman researcher who was lamenting the EDI time commitments that she was facing as a scientist, woman, and diverse scholar. We must do better. Women and diverse scholars have been ringing the alarm bell for decades and the allyship and workload must be taken on by all of us. We can't address unfairness by placing more demands on those who already face multiple injustices.

While I pointed out to that young woman how much had changed since I was an early career researcher—we at least have EDI offices and discussions on inclusion—I sympathized with her plight. I told her that my hope, going forward, is that we'll start to see institutions act on the big-ticket items: tenure, full professorships, equal pay, ending abuse, discrimination, and harassment, a better culture, and tangible opportunities for her and for all women at the university. And I shared that extraordinary, inspiring, and smart women like her have been speaking up for almost two centuries and that she—and indeed, all of us—has that same opportunity to lead and to make her mark to better universities, colleges, and Canada.

We need more women leaders. Women have always had the "right stuff." They know their value. They think big, act strongly, and take risks. They are bold, determined, and disciplined, and they do not need to be told "to fake it till they make it" or to be less apologetic. But as I write these last sentences, the United Nations is warning that equality among the sexes is slipping before our eyes and remains 300 years away. Until we have a more equal world, we all need to stand and be counted in support of women and their rights, and in support of an end to the sexist and racist legacy in science and research and in our institutions.

Acknowledgements

I AM GRATEFUL FOR the generations of courageous, trail-blazing women who came before us—women who fought for university education, higher degrees, professorships, and leadership positions. Women who choose science and give us their important perspectives, and women who have repeatedly pointed out injustice and continue to fight for equality and their rightful place in STEM fields. Women who share their stories and uncover history. In short, women who dare.

I have done my best to honor each of the researchers and authors covered recognizing here, their contributions and providing their proper titles. This matters as far too often women's names and contributions have been dropped from recorded history. I am grateful for every conversation with women researchers about their successes, hopes, and challenges. What a privilege to learn from each of them!

Thank you to the University of Toronto Press for their work in bringing "Hunting the 1918 Flu: One Scientist's Search for a Killer Virus" to life in 2003, and for generously allowing me to excerpt the book in chapter one here.

Thank you to Dr. Sven Spengemann who read draft after draft of every chapter and who is a tireless ally for women.

ACKNOWLEDGEMENTS

Professor Kevin Smith for every conversation about science and medicine, Professor Antonia Chayes for her inspiring stories and kind support, Professor Jeremy Kerr for his advice over a decade, review of chapters, and career-long support of women researchers, Professor Gina Rippon for her important research and kind review of a number of chapters, and Drs Sandhya and Swapna Mylabathula for their review and comments. Professor Imogen Coe, Professor Liz Grant, Dr. Andy Kerr, Dr. Simona Marinescu, Professor Allison Sekuler, Professor Malinda Smith, Professor Malik Dahlan and Sarah Yamani and their beautiful family, Deborah Gordon-El-Bihbety, and Heba Jibril for the numerous conversations to build a better world.

Thanks to Kerrie Kennedy for every conversation about women in science, all of them in Scottish Gaelic. And to Yvonne Nichols, Glenna Mackay-Johnstone, Holly Tomilson, Stephanie Casson, and Shirley Macmillan for every conversation about women, a lifetime of friendship, and unending support this year. Also, the tremendous women with whom I work, and the doctors, nurses, and healthcare providers for their science, caring, and humanity. And my parents, Helen and Errol Duncan, for their love, support, and unending belief in an equal and just world, as well as the Spengemann family, Jeannette Marseu, Christopher Duncan, Cynthia Miller, and their families.

Finally, I am grateful to Kenneth Whyte, a giant in Canadian journalism, for his belief in this book and his extraordinary editing skills, and the outstanding team at Sutherland House. It has been an honor, a pleasure, and a joy.

References

AAU, "AAU President Reiterates the Value of Diverse Campus Communities," AAU. June, 29, 2023. https://www.aau.edu/newsroom/press-releases/aau-president-reiterates-value-diverse-campus-communities.

AAUW, *Barriers and Bias: The Status of Women in Leadership* (Washington DC, AAUW, 2016).

AAUW, *Why So Few? Women in Science, Technology, Engineering, and Mathematics* (Washington DC, AAUW, 2010).

AAUW, "The Trouble with Tracking," AAUW. No date. https://www.aauw.org/issues/education/gender-bias/.

Accenture, "Accenture Finds Girls' Take-up of STEM Subjects is Held Back by Stereotypes, Negative Perceptions and Poor Understanding of Career Options," Accenture. February, 7, 2017. https://newsroom.accenture.com/news/accenture-finds-girls-take-up-of-stem-subjects-is-held-back-by-stereotypes-negative-perceptions-and-poor-understanding-of-career-options.htm.

Admin, "Types of Violence Against Women and Girls," UN Women Australia. November, 24, 2020. https://unwomen.org.au/types-of-violence-against-women-and-girls/.

Alam, A., "Mapping Gender Equality in STEM from School to Work. The Story of Disparities in STEM in 17 Illuminating Charts," UNICEF. November, 20, 2020. https://www.unicef.org/globalinsight/stories/mapping-gender-equality-stem-school-work.

American Institute of Physics, "Women in Physics and Astronomy, 2019," AIP. March, 2019. https://www.aip.org/statistics/reports/women-physics-and-astronomy-2019.

Angier, N., "Insights from the Youngest Minds," *The New York Times*. April, 30, 2012. https://www.nytimes.com/2012/05/01/science/insights-in-human-knowledge-from-the-minds-of-babes.html.

REFERENCES

Anonymous, "Queen's School of Medicine: Confronting Exclusion," Queen's University. July, 7, 2020a. https://www.queensu.ca/alumnireview/articles/ 2020-07-17/queen-s-school-of-medicine-confronting-exclusion#:~:text= In%201918%2C%20a%20motion%20to,Edward%20Thomas%20(Cultural %20Studies).

Anonymous, "Systemic Racism: Science Must Listen, Learn and Change," *Nature*. June, 9, 2020b. https://www.nature.com/articles/d41586-020-01678-x.

Anonymous, "40 Years of Women at Hart House," Hart House. May, 29, 2019. https://harthouse.ca/blog/40-years-of-women-at-hart-house.

Anonymous, "The Nobel Gender Gap Is Worse Than You Think," *Nature Index*. October, 8, 2018. https://www.nature.com/nature-index/news-blog/ the-nobel-gender-gap-is-worse-than-you-think.

Anonymous, "Women in Science: Women's work." *Nature*, 495, 7439 (2013: 21).

Anonymous, "Death of Evidence." *Nature* 487, 7407 (2012: 271–272).

Asare, J., "How to Increase Female Representation in the STEM Field," *Forbes*. October, 16, 2018. https://www.forbes.com/sites/janicegassam/2018/10/ 16/how-to-increase-female-representation-in-the-stem-field/?sh=68c9ad 11466a.

Asmelash, L., "In 5 Years of #MeToo, Here's What's Changed—and What Hasn't," CNN. October, 27, 2022. https://www.cnn.com/2022/10/27/us/ metoo-five-years-later-cec/index.html.

Asmelash, L., "Woman Who Popularized the Gender Reveal Party Says Enough Already After Latest Wildfire," CNN. September, 7, 2020. https://www.cnn. com/2020/09/07/us/gender-reveal-parties-overview-trnd/index.html#: ~:text=This%20week%2C%20as%20more%20than,Stop%20having%20 these%20stupid%20parties.

Aubourg, L., "Girls More Likely to Attribute Failure to Lack of Talent: Study," PhysOrg. March, 10, 2022. https://phys.org/news/2022-03-girls-attribute- failure-lack-talent.html.

Bailey, A., Williams, A., and Cimpian, A., "Based on Billions of Words on the Internet, PEOPLE = MEN." *Science*, 8, 13 (2022).

Baker, M., Halberstam, Y., Kroft, K. et al., "Pay Transparency and the Gender Gap," Statistics Canada. September 16, 2019. https://www150.statcan. gc.ca/n1/pub/11f0019m/11f0019m2019018-eng.htm.

Baldwin, M., "How 'Man of Science' Was Dumped in Favor of 'Scientist'," The Conversation. August, 5, 2014. https://theconversation.com/how-man-of- science-was-dumped-in-favour-of-scientist-30132.

Ball, P., "Science Hasn't Gone 'Woke'—The Only People Meddling with It Are the Tories," *The Guardian*. October, 4, 2023. https://www.theguardian. com/commentisfree/2023/oct/04/science-research-sex-gender-michelle-donelan-conservatives-conference.

Barnes, B., "Eminent Archaeologist Mary Leakey Dies at 83," *The Washington Post*. December, 10, 1996. https://www.washingtonpost.com/archive/ local/1996/12/10/eminent-archaeologist-mary-leakey-dies-at-83/ d2010875-7d07-42fb-9532-c237405940d0/?utm_term=.5705568a41a0. %2022/10/2022.

Barnett, R. and Rivers, C., "We've Studied Gender and STEM for 25 Years. The Science Doesn't Support the Google Memo," VOX. April, 11, 2017. https://www.vox.com/2017/8/11/16127992/google-engineer-memo-research-science-women-biology-tech-james-damore.

Baron-Cohen, S. *The Pattern Seekers: How Autism Drives Human Invention* (New York, Basic Books, 2023).

Baron-Cohen, S., Knickmeyer, R., and Belmonte, M. "Sex Differences in the Brain: Implications for Explaining Autism." *Science*, 310, 5749 (2005: 819–823).

Baston, J., *Her Oxford* (Nashville, Vanderbilt University Press, 2008).

Batty, D. and Davis, N., "Why Science Breeds a Culture of Sexism," *The Guardian*. July, 7, 2018. https://www.theguardian.com/science/2018/jul/07/ why-science-is-breeding-ground-for-sexism.

Bayer Corporation, "Bayer Facts of Science Education XV: A View from the Gatekeepers—STEM Department Chairs at America's Top 200 Research Universities on Female and Underrepresented Minority Undergraduate STEM Students." *Journal of Science Education and Technology*, 21, 3 (2012: 317–324).

Bayer Corporation, "Bayer Facts of Science Education XIV: Female and Minority Chemists and Chemical Engineers Speak About Diversity and Underrepresentation in STEM," Bayer. 2010. https://www.bayer.com/sites/ default/files/2010_fose_xiv_0.pdf.

BBC, "Harvey Weinstein Timeline: How the Scandal Has Unfolded," BBC. February, 24, 2023. https://www.bbc.com/news/entertainment-arts-41594672.

BBC, "Nobel Prize: We Will Not Have Gender or Ethnicity Quotas - Top Scientist," BBC. October, 11, 2021. https://www.bbc.com/news/world-europe-58875152.

BBC, "Cambridge University Marks 150 Years of Female Students," BBC. September, 2019a. https://www.bbc.com/news/uk-england-cambridgeshire-49595057.

BBC, "James Watson: Scientist Loses Titles After Claims Over Race," BBC. January, 13, 2019b. https://www.bbc.com/news/world-us-canada-46856779.

BBC, "Female Scientists Post 'Distractingly Sexy' Photos," BBC. June, 11, 2015. https://www.bbc.com/news/blogs-trending-33099289.

Becker, K. and Miller, L., "Preparing for the Tenure-track Interview," *University Affairs*. February, 11, 2015. https://www.universityaffairs.ca/career-advice/career-advice-article/preparing-tenure-track-interview/.

Bell, D., "Our Histories Are Complicated: Famous Five Fought a Good but Imperfect Fight," CBC. October, 18, 2019. https://www.cbc.ca/news/canada/calgary/famous-five-fought-good-imperfect-fight-1.5325290#:~:text=Calgary%C2%B7Q%26A-,'Our%20histories%20are%20complicated'%3A%20Famous%20Five%20fought%20a%20good,Five%20%E2%80%94%20led%20to%20Persons%20Day.

Bello, A., Blowers, T., and Schneegans, S., "To Be Smart, the Digital Revolution will Need to be Inclusive," In: *UNESCO Science Report: The Race against Time for Smarter Development* (Paris, UNESCO, 2021).

Bernstein, R., "Science Still Seen as Male Profession, According to International Study of Gender Bias," *Science*. May, 22, 2015. https://www.science.org/content/article/science-still-seen-male-profession-according-international-study-gender-bias.

Beveridge, W., *Influenza: The Last Great Plague. An Unfinished Story of Discovery* (New York, Prodist, 1977).

Bian. L., Leslie, S., and Cimpian, A., "Evidence of Bias Against Girls and Women in Contexts that Emphasize Intellectual Ability." *American Psychologist*, 73, 9 (2018a: 1139–1153).

Bian, L., Leslie, S., Murphy, M. et al., "Messages about Brilliance Undermine Women's Interest in Educational and Professional Opportunities." *Journal of Experimental Social Psychology*, 76 (2018b: 404–420).

Bian, L., Leslie, S., and Cimpian, A., "Gender Stereotypes About Intellectual Ability Emerge Early and Influence Children's Interests." *Science*, 355, 6323 (2017: 389–391).

Bierschbach, B., "This Woman Fought to End Minnesota's 'Marital Rape' Exception, and Won," NPR. May, 4, 2019. https://www.npr.org/2019/05/04/719635969/this-woman-fought-to-end-minnesotas-marital-rape-exception-and-won.

Binns, C., "What's Behind the Pay Gap in STEM Jobs?" *Stanford Business*. February, 19, 2021. https://www.gsb.stanford.edu/insights/whats-behind-pay-gap-stem-jobs.

Booker, B., "Princeton University Agrees to Nearly $1 Million in Back Pay to Female Professors," NPR. October, 13, 2020. https://www.npr.org/2020/10/13/923233877/princeton-university-agrees-to-nearly-1-million-in-back-pay-to-female-professors.

Bowling, J. and Martin, B., "Science: A Masculine Disorder?" *Science and Public Policy*, 12, 6, (1985: 308–316).

Brandt, S., Cotner, S., Koth, Z. et al., "Scientist Spotlights: Online Assignments to Promote Inclusion in Ecology and Evolution." *Ecology and Evolution*, 10, 22 (2020: 12450–12456).

Bray, S., Alcalde, M., and Subramaniam, M., "Women in Leadership: Challenges and Recommendations," *Inside Higher Education*. July, 17, 2020. https://www.insidehighered.com/views/2020/07/17/women-leadership-academe-still-face-challenges-structures-systems-and-mind-sets.

British Science Council, "How Do We Define a Scientist?" Science Council. June, 24, 2016. https://sciencecouncil.org/how-do-we-define-a-scientist/.

Brown, B., "Toronto Professor's in Hot Water for Underwater 'Leering' in Pool," *The Buffalo News*. April, 2, 1989.

Budday, S., Steinmann, P., and Kuhl, E., "Physical Biology of Human Brain Development." *Frontiers in Cellular Neuroscience*, 9, 257 (2015: 1–17).

Bueckert, K., "All Female Faculty at University of Waterloo Get Raises After Gender Wage Gap Discovered," CBC. August, 8, 2016. https://www.cbc.ca/news/canada/kitchener-waterloo/university-waterloo-female-faculty-wage-gap-raises-1.3711694.

Buolamwini, J., "Artificial Intelligence Has a Problem with Gender and Racial Bias. Here's How to Solve It," *Time*. February, 7, 2019. https://time.com/5520558/artificial-intelligence-racial-gender-bias/.

Burczycka, M., "Students' Experiences of Unwanted Sexualized Behaviours and Sexual Assault at Postsecondary Schools in the Canadian Provinces, 2019," Statistics Canada. September, 14, 2020. https://www150.statcan.gc.ca/n1/pub/85-002-x/2020001/article/00005-eng.htm.

Burgen, S., "Catalonia to Pardon Up to 1,000 People Accused of Witchcraft." *The Guardian*. January, 26, 2022. https://www.theguardian.com/world/2022/jan/26/catalonia-expected-to-pardon-up-to-1000-people-accused-of-witchcraft.

Burnell, J., "Me," University of Glasgow. April, 2023. https://www.gla.ac.uk/explore/avenue/me/mebyjocelynbellburnell/.

Busch, M., "8 Tips to Develop Children's Curiosity," Mayo Clinic Health System.

May,15, 2020. https://www.mayoclinichealthsystem.org/hometown-health/speaking-of-health/8-tips-to-develop-childrens-curiosity.

Bushwick, S., Harper, K., and DelViscio, J. "Scientists Are Beginning to Learn the Language of Bats and Bees Using AI." *Scientific American.* September, 11, 2023. https://www.scientificamerican.com/podcast/episode/scientists-are-beginning-to-learn-the-language-of-bats-and-bees-using-ai/.

Byars-Winston, A. and M. Dahlberg (Eds), *The Science of Effective Mentorship in STEMM* (Washington DC, The National Academies Press, 2019).

Cadloff, E., "What Happens to Sexual Assault Reports at Canadian Universities? No One Really Knows," *Macleans.* November, 15, 2022. https://education.macleans.ca/feature/what-happens-to-sexual-assault-reports-at-canadian-universities-no-one-really-knows/.

Cafley, J., "The Struggle for Gender Equity in University Leadership," University Affairs. October, 27, 2021. https://www.universityaffairs.ca/features/feature-article/the-struggle-for-gender-equity-in-university-leadership/.

Calaza, K., Erthal, F., Pereira, M. et al., "Facing Racism and Sexism in Science by Fighting Against Social Implicit Bias: A Latina and Black Woman's Perspective," *Frontiers in Psychology*, 12, 671481 (2021: 1–9).

Calisi, R. and a Working Group of Mothers in Science, "How to Tackle the Childcare–Conference Conundrum." *Proceedings of the National Academies of Sciences of the United States of America*, 115, 12 (2018: 2845–2849).

Callier, V., "The Complex Role of Gender in Faculty Hiring," *Science.* April, 15, 2016. https://www.science.org/content/article/complex-role-gender-faculty-hiring.

Campbell, N., "The Word 'Scientist' or its Substitute." *Nature*, 114 (1924: 788).

Canada's Aviation Hall of Fame, "Elizabeth Muriel Gregory MacGill," CAHF. 2023. https://cahf.ca/elizabeth-muriel-gregory-macgill/.

Cantor, D., Fisher, B., Chibnall, S. et al., *Report on the AAU Campus Climate Survey on Sexual Assault and Misconduct* (Maryland, Westat, 2020).

Caranci, B., Judge, K., and Kobelak, O., "Women and STEM: Bridging the Divide," TD. September, 12, 2017. http://economics.td.com/women-and-stem-bridging-divide.

Carstairs, C. and Hughes, K., "The Long Fight Against Sexual Assault and Harassment at Universities," The Conversation. December, 5, 2021. https://theconversation.com/the-long-fight-against-sexual-assault-and-harassment-at-universities-170258.

Casarez, J., Perez, E., and Frehse, R., "Olympic Gymnasts Biles, Raisman and

Maroney Are Among Dozens Seeking $1 Billion from the FBI Over Botched Larry Nassar Sex Abuse Investigation," CNN. June, 9, 2022. https://www.cnn.com/2022/06/08/us/larry-nassar-fbi-mishandling-claim/index.html.

Casey, M., "Claudine Gay to Be Harvard's 1st Black President, 2nd Woman," AP News. December, 15, 2022. https://apnews.com/article/education-penny-pritzker-08ae6297a31f4f42b85a2813336b33e2.

Catudella, J., "When Women Came to Queen's." *Canadian Medical Association Journal*, 161, 5 (1999: 575–576).

CBC, "Women's Brains ARE Built for Science. Modern Neuroscience Explodes an Old Myth," CBC. September, 20, 2019. https://www.cbc.ca/radio/quirks/july-25-2020-women-in-science-special-how-science-has-done-women-wrong-1.5291077/women-s-brains-are-built-for-science-modern-neuroscience-explodes-an-old-myth-1.5291081.

CBC, "Right Stuff, Wrong Gender—the True Story of the Women Who Almost Went to the Moon," CBC. April, 27, 2018. https://www.cbc.ca/radio/quirks/scrap-medical-weed-women-in-space-and-more-1.4636793/right-stuff-wrong-gender-the-true-story-of-the-women-who-almost-went-to-the-moon-1.4636802.

CBC, "Canadian Elsie MacGill Was the First Female Aeronautical Engineer in the World," CBC. May, 5, 2017. https://www.cbc.ca/2017/canadathestoryofus/canadian-elsie-macgill-was-the-first-female-aeronautical-engineer-in-the-world-1.4099967.

CBC News, "Laurier Raises Pay for Female Profs After Gender Equity Analysis," CBC. May, 10, 2017. https://www.cbc.ca/news/canada/kitchener-waterloo/wilfrid-laurier-university-gender-wage-gap-pay-raise-1.4108256.

CBC News, "Female McMaster Professors Getting a Pay Boost to Same Level as Men," CBC. April, 28, 2015. https://www.cbc.ca/news/canada/hamilton/headlines/female-mcmaster-professors-getting-a-pay-boost-to-same-level-as-men-1.3052626.

Cech, E. and Blair-Loy, M., "The Changing Career Trajectories of New Parents in STEM." *Proceedings of the National Academy of Sciences*, 116, 10 (2019: 4182–4187).

Center on the Developing Child, "In Brief: The Science of Early Childhood Development," Center on Developing Child. 2008. www.developingchild.harvard.edu.

Chapin, A., "Four Decades After the Battle of the Sexes, the Fight for Equality

Goes On," *The Guardian*. March, 11, 2017. https://www.theguardian.com/sport/2017/mar/11/billie-jean-king-battle-of-the-sexes-tennis.

Charlesworth, T. and Banaji, M., "Gender in Science, Technology, Engineering, and Mathematics: Issues, Causes, Solutions." *Journal of Neurosci*ence, 39, 37 (2019: 7228–7243).

Chung, E., "How Engineers of the Montreal Massacre Generation are Changing the World," CBC. December, 6, 2019a. https://www.cbc.ca/news/science/montreal-massacre-women-engineer-profiles-1.5385088.

Chung, E., "Half of Canadians Can't Name a Woman Scientist or Engineer, Poll Finds," CBC. March, 8, 2019b. https://www.cbc.ca/news/science/women-scientists-1.5048491.

Clancy, K., Lee, K., Rodgers, E. et al., "Double Jeopardy in Astronomy and Planetary Science: Women of Color Face Greater Risks of Gendered and Racial Harassment." *Journal of Geophysical Research: Planets*, 122, 7 (2017:1610–1623).

Clancy, K., Nelson, R., and Rutherford, J., "Survey of Academic Field Experiences (SAFE): Trainees Report Harassment and Assault." PLoS ONE. July, 16, 2014. https://journals.plos.org/plosone/article?id=10.1371/journal.pone.0102172.

Clayton, J. and Collins, F., "Policy: NIH to Balance Sex in Cell and Animal Studies." *Nature*, 509 (2014: 282–283).

Cleghorn, E., *Unwell Women: Misdiagnosis and Myth in a Man-made World* (New York, Dutton, 2021).

CNN, "How Baby Brains Develop," CNN. September, 25, 2014. https://www.cnn.com/videos/health/2014/09/25/sgmd-gupta-baby-brain-development.cnn.

Cohen, A., *Imbeciles: The Supreme Court, American Eugenics, and The Sterilization of Carrie Buck* (New York, Penguin Books, 2016).

Cohen, S., "The Day Women Went on Strike," *Time*. August, 26, 2015. https://time.com/4008060/women-strike-equality-1970/.

CohenMiller, A. and Izekenova, Z., "Motherhood in Academia During the COVID-19 Pandemic: An International Online Photovoice Study Addressing Issues of Equity and Inclusion in Higher Education." *Innovative Higher Education*, 47, (2022: 813–835).

Cohn, D., "It's Official: Minority Babies are the Majority Among the Nation's Infants, But Only Just," Pew Research Center. June, 23, 2016. https://www.pewresearch.org/short-reads/2016/06/23/its-official-minority-babies-are-the-majority-among-the-nations-infants-but-only-just/.

Cole, N., "Understanding Socialization in Sociology: Overview and Discussion of a Key Sociological Concept," ThoughtCo. January, 30, 2020. https://www.thoughtco.com/socialization-in-sociology-4104466.

College of Medicine and Veterinary Medicine, "Sophia Jex-Blake and the Edinburgh Seven," The University of Edinburgh. January, 8, 2018. https://www.ed.ac.uk/medicine-vet-medicine/about/history/women/sophia-jex-blake-and-the-edinburgh-seven. 27/11/2022.

Collier, R., *The Plague of the Spanish Lady: The Influenza Pandemic of 1918–1919* (New York, Athenum, 1974).

Collins, C., Landivar, L., and Ruppanner, L., "COVID-19 and the Gender Gap in Work Hours." *Gender, Work & Organization*, 28, S1 (2021: 101–112).

Colwell, R., "Women Scientists Have the Evidence About Sexism," *The Atlantic*. August, 30, 2020. https://www.theatlantic.com/ideas/archive/2020/08/women-scientists-have-evidence-about-sexism-science/615823/.

Commodore, F., "6 of 8 Ivy Leagues Will Soon Have Women as Presidents—An Expert Explains Why This Matters," The Conversation. April, 4, 2023. https://theconversation.com/6-of-8-ivy-leagues-will-soon-have-women-as-presidents-an-expert-explains-why-this-matters-201821.

Common Sense Media, "Watching Gender: How Stereotypes in Movies and on TV Impact Kids' Development," Common Sense. June, 19, 2017. https://www.commonsensemedia.org/sites/default/files/research/report/2017_commonsense_watchinggender_fullreport_0620.pdf.

Cooke, E., "Why Women Aren't from Venus, and Men Aren't from Mars." *Nature*. November, 18, 2022. https://www.nature.com/articles/d41586-022-03782-6.

Corbyn, Z., "BethAnn McLaughlin: 'Too Many Women in Science Have to Run the Gauntlet of Abuse and Leave," *The Guardian*. April, 7, 2019. https://www.theguardian.com/science/2019/apr/07/bethann-mclaughlin-sexual-harassment-in-science.

Coughlan, S., "Oxford University First Female Head," BBC News. May, 28, 2015. https://www.bbc.com/news/education-32916421.

Council of Canadian Academies, *Degrees of Success: The Expert Panel on the Labour Market Transition of PhD Graduates* (Ottawa, Council of Canadian Academies, 2021).

Cox, J., "The Trust Crisis Facing Women leaders," BBC. November, 30, 2022. https://www.bbc.com/worklife/article/20221129-the-trust-crisis-facing-women-leaders.

Criado Perez, C., *Invisible Women: Exposing Data Bias in a World Designed for Men* (New York, Harry N. Abrams, 2019).

REFERENCES

Criado Perez, C., "The Deadly Truth About a World Built for Men–from Stab Vests to Car Crashes." *The Guardian*. February, 23, 2019. https://www.theguardian.com/lifeandstyle/2019/feb/23/truth-world-built-for-men-car-crashes.

Crim, K., "Notes on the Intelligence of Women. Atlantic Authors from the Early to the Late Twentieth Century Comment on the Status of Women in Science," *The Atlantic*. May, 2005. https://www.theatlantic.com/magazine/archive/2005/05/notes-on-the-intelligence-of-women/304042/.

Crosby, A., *Epidemic and Peace, 1918* (Westport, Greenwood Press, 1976).

Cummings, M., "Pay Gap Between Male and Female Professors Continues to Plague Canadian Universities," CBC. September, 28, 2020. https://www.cbc.ca/news/canada/edmonton/gender-pay-gap-persists-at-canadian-universities-1.5739466.

D'Mello, A., "What Is the Social Brain?" McGovern Institute. October 4, 2019. https://mcgovern.mit.edu/2019/10/04/what-is-the-social-brain/.

Darby, N., *A History of Women's Lives in Oxford* (Barnsley, Pen and Sword, 2019).

Davis, P., Meagher, E., Pomeroy, C. et al., "Pandemic-related Barriers to the Success of Women in Research: A Framework for Action." *Nature Medicine*, 28 (2022: 436–438).

Dejevsky, M., "The First Woman in Space: 'People Shouldn't Waste Money on Wars,'" *The Guardian*. March, 29, 2017. https://www.theguardian.com/global-development-professionals-network/2017/mar/29/valentina-tereshkova-first-woman-in-space-people-waste-money-on-wars.

Deloitte AI Institute, "Women in AI," Deloitte. September, 30, 2020. https://www2.deloitte.com/content/dam/Deloitte/us/Documents/deloitte-analytics/us-consulting-women-in-ai.pdf.

Deloitte., "Understanding the Pandemic's Impact on Working Women. How Employers Can Act Now to Prevent a Setback in Achieving Gender Parity in the Workplace," Deloitte. 2020. https://www.deloitte.com/content/dam/assets-shared/legacy/docs/about/2022/gx-about-deloitte-understanding-the-pandemic-s-impact-on-working-women.pdf.

Des Jardins, J., "Madame Curie's Passion," *Smithsonian*. October, 2011. https://www.smithsonianmag.com/history/madame-curies-passion-74183598/.

Deslauriers, M., "Sexual Difference in Aristotle's Politics and His Biology." *Classical World*, 102, 3, (2009: 215–231).

Deveau, D., "A Lifetime of Courage: Anne Innis Dagg," *Financial Post*. November,

17, 2022. https://financialpost.com/executive/executive-women/a-lifetime-of-courage-anne-innis-dagg.

Diaz, D., "3 Times Trump Defended his 'Locker Room' Talk," CNN. October, 9, 2016. https://www.cnn.com/2016/10/09/politics/donald-trump-locker-room-talk-presidential-debate-2016-election/index.html.

Dominus, S., "Women Scientists Were Written Out of History. It's Margaret Rossiter's Lifelong Mission to Fix That," *Smithsonian*. October, 2019. https://www.smithsonianmag.com/science-nature/unheralded-women-scientists-finally-getting-their-due-180973082/.

Donald, A., "67% of Europeans Don't Believe Women Have the Skills to Be Scientists," *The Guardian*. September, 24, 2015. https://www.theguardian.com/women-in-leadership/2015/sep/24/67-of-europeans-dont-believe-women-have-the-skills-to-be-scientists.

Dubé, D. and Vuchnich, A., "Heart Disease in Women Is Under-diagnosed, Under-treated and Under-researched: Heart and Stroke Report," Global News. February, 1, 2018. https://globalnews.ca/news/3998811/heart-disease-in-women-is-under-diagnosed-under-treated-and-under-researched-heart-and-stroke-report/.

Duke, S., "Will AI Make the Gender Gap in the Workplace Harder to Close?" World Economic Forum. December, 21, 2018. https://www.weforum.org/agenda/2018/12/artificial-intelligence-ai-gender-gap-workplace/.

Duncan, K., *Hunting the 1918 Flu: One Scientist's Search for a Killer Virus* (Toronto, University of Toronto Press, 2003).

Dutt, K., Pfaff, D., Bernstein, A. et al., "Gender Differences in Recommendation Letters for Postdoctoral Fellowships in Geoscience." *Nature Geoscience*, 9 (2016: 805–808).

Edge, L., "Science Has a Racism Problem." *Cell*, 181 (2020: 1443–1444).

Editorial, "The Cooperative Human." *Nature Human Behaviour*, 2 (2018: 427–428).

Eisenberger, N. and Lieberman, M., "Why It Hurts to Be Left Out: The Neurocognitive Overlap Between Physical and Social Pain." In: Williams, K., Forgas, J., von Hippel, W. (eds). *The Social Outcast: Ostracism, Social Exclusion, Rejection, and Bullying* (London, Psychology Press, 2005).

Eisenberger, N., Lieberman, M., and Williams, K., "Does Rejection Hurt? An fMRI Study of Social Exclusion." *Science*, 302, 5643 (2003: 290–292).

Elections Canada, "A Brief History of Federal Voting Rights in Canada," Elections Canada. 2023. https://electionsanddemocracy.ca/voting-rights-through-time-0/brief-history-federal-voting-rights-canada.

Eliot, L., Ahmed, A., Khan, H. et al., "Dump the 'Dimorphism': Comprehensive Synthesis of Human Brain Studies Reveals Few Male-Female Differences Beyond Size." *Neuroscience & Biobehavioral Reviews*, 125 (2021: 667–697).

Eliot, L., "Bad Science and the Unisex Brain." *Nature*, 566, 7455 (2019: 453–454).

Eliot, L., "The Unisex Brain." Rosalind Franklin University. November, 19, 2018. https://www.rosalindfranklin.edu/news/the-unisex-brain/.

Eliot, L., *Pink Brain, Blue Brain: How Small Differences Grow into Troublesome Gaps and What We Can Do About It* (New York, Houghton Mifflin Harcourt, 2009).

Elsworthy, E., "Curious Children Ask 73 Questions Each Day - Many of Which Parents Can't Answer, Says Study." *Independent*. December, 3, 2017. https://www.independent.co.uk/news/uk/home-news/curious-children-questions-parenting-mum-dad-google-answers-inquisitive-argos-toddlers-chad-valley-tots-town-a8089821.html.

Engineering, "Women in Engineering at U of T: A Timeline," University of Toronto Engineering News. January, 28, 2015. https://news.engineering.utoronto.ca/women-engineering-u-t-timeline/.

European Institute for Gender Equality, "Economic Benefits of Gender Equality in the EU. How Gender Equality in STEM Education Leads to Economic Growth," EIGE. August, 10, 2017. https://eige.europa.eu/publications-resources/publications/economic-benefits-gender-equality-eu-how-gender-equality-stem-education-leads-economic-growth?language_content_entity=en.

Eveleth, R., "Soviet Russia Had a Better Record of Training Women in STEM Than America Does Today," *Smithsonian*. December, 12, 2013. https://www.smithsonianmag.com/smart-news/soviet-russia-had-a-better-record-of-training-women-in-stem-than-america-does-today-180948141/.

Falk, S., *The Light Ages: The Surprising Story of Medieval Science* (New York, W. W. Norton, 2020).

Famous5 Foundation, "The 'Persons' Case," Famous5 Foundation. No date. https://www.famous5.ca/the-persons-case.

Farley, L., "I Coined the Term 'Sexual Harassment.' Corporations Stole It," *The New York Times*. October, 18, 2017. https://www.nytimes.com/2017/10/18/opinion/sexual-harassment-corporations-steal.html.

Fathima, F., Awor, P., Yi-Chun, Y. et al., "Challenges and Coping Strategies Faced by Female Scientists—A Multicentric Cross Sectional Study," PLoS ONE. September, 21, 2020. https://journals.plos.org/plosone/article?id=10.1371/journal.pone.0238635.

Ferguson, D., "'Why Are They Not on Wikipedia?': Dr. Jess Wade's Mission for Recognition for Unsung Scientists," *The Guardian*. October, 1, 2023. https://www.theguardian.com/science/2023/oct/01/why-are-they-not-on-wikipedia-dr-jess-wades-mission-for-recognition-for-unsung-scientists.

Firth-Butterfield, K. and Ammanath, B., "5 Ways to Get More Women Working in AI," World Economic Forum, August, 12, 2021. https://www.weforum.org/agenda/2021/08/5-ways-increase-women-working-ai/.

Fitzpatrick, M., "Death of Scientific Evidence Mourned on Parliament Hill," CBC. July, 10, 2012. https://www.cbc.ca/news/politics/death-of-scientific-evidence-mourned-on-parliament-hill-1.1218019.

Focquaert, F., Steven, M., Wolford, G. et al., "Empathizing and Systemizing Cognitive Traits in the Sciences and Humanities." *Personality and Individual Differences*, 43, 3 (2007: 619–625).

Ford, A., *A Path Not Strewn with Roses: One Hundred Years of Women at the University of Toronto 1884–1984* (Toronto, University of Toronto Press, 1985).

Fouad, N., Chang, W., Singh, M. et al., "Women's Reasons for Leaving the Engineering Field." *Frontiers in Psychology*, 8, 875 (2017).

Fox, G. "Meet the Neuroscientist Shattering the Myth of the Gendered Brain." *The Guardian*. February, 24, 2019. https://www.theguardian.com/science/2019/feb/24/meet-the-neuroscientist-shattering-the-myth-of-the-gendered-brain-gina-rippon.

France-Presse, A., "Nobel Prize Will Have No Gender or Ethnicity Quotas, Academy Head Says," *The Guardian*. October, 12, 2021. https://www.theguardian.com/science/2021/oct/12/nobel-prize-will-have-no-gender-or-ethnicity-quotas-academy-head-says.

Frank, K., "A Gender Analysis of the Occupational Pathways of STEM Graduates in Canada," Statistics Canada. September, 16, 2019. https://www150.statcan.gc.ca/n1/pub/11f0019m/11f0019m2019017-eng.htm.

Fromson, D., "FDR Grew Up in a Dress: It Wasn't Always Blue for Boys and Pink for Girls," *The Atlantic*. April, 14, 2011. https://www.theatlantic.com/national/archive/2011/04/fdr-grew-up-in-a-dress-it-wasnt-always-blue-for-boys-and-pink-for-girls/237299/.

Fry, R., Kennedy, B., and Funk, C., "STEM Jobs See Uneven Progress in Increasing Gender, Racial and Ethnic Diversity," Pew Research Center. April, 1, 2021. https://www.pewresearch.org/science/2021/04/01/stem-jobs-see-uneven-progress-in-increasing-gender-racial-and-ethnic-diversity/.

Funk, C. and Parker, K., *Women and Men in STEM Often at Odds Over Workplace Equity* (Washington DC, Pew Research Center, 2018).

Funk, W., "Mercury 13," Ninety-nines. June, 1, 2009. https://www.ninety-nines. org/mercury13.htm.

Gagliardi, J., Espinosa, L., Turk, J. et al., *American College President Study 2017* (Washington DC, American Council on Education, 2017).

George, W., "Notes on the Intelligence of Woman," *The Atlantic.* December, 1915. https://www.theatlantic.com/magazine/archive/1915/12/notes-on-the-intelligence-of-woman/304038/.

Gibney, E., "What the Nobels Are—and Aren't—Doing to Encourage Diversity." *Nature*, 562, 19 (2018).

Gjersoe, N., "Bridging the Gender Gap: Why Do So Few Girls Study STEM Subjects," *The Guardian.* March, 8, 2018. https://www.theguardian.com/science/head-quarters/2018/mar/08/bridging-the-gender-gap-why-do-so-few-girls-study-stem-subjects.

Gladwell, M., "Getting In," *The New Yorker.* October, 2, 2005. https://www.newyorker. com/magazine/2005/10/10/getting-in-ivy-league-college-admissions.

Goldenberg, S., "Why Women Are Poor at Science, by Harvard President," *The Guardian.* January, 18, 2005. https://www.theguardian.com/science/2005/jan/18/educationsgendergap.genderissues#:~:text=The%20president%20of%20Harvard%20University,career%20barrier%20for%20female%20academics.

Gopnik, A., "Scientific Thinking in Young Children: Theoretical Advances, Empirical Research, and Policy Implications." *Science*, 337, 6102 (2012: 1623–1627).

Gopnik, A. "What Do Babies Think?" TED. July, 2011. https://www.ted.com/talks/alison_gopnik_what_do_babies_think?language=en.

Gopnik, A., Meltzoff, A., and Kuhl, P., *The Scientist in the Crib. What Early Learning Tells Us About the Mind* (New York, Perennial, 2000).

Government of Canada., "Women at War," Veterans Affairs Canada. October, 30, 2023a. https://www.veterans.gc.ca/eng/remembrance/classroom/fact-sheets/women.

Government of Canada, "Persons' Day," Women and Gender Equality Canada. September, 14, 2023b. https://women-gender-equality.canada.ca/en/commemorations-celebrations/womens-history-month/persons-day.html.

Government of Canada, "Elsie MacGill," Veterans Affairs Canada. January, 22, 2020. https://www.veterans.gc.ca/eng/remembrance/people-and-stories/elsie-macgill.

Government of Canada, "The Battle of Britain: Fighting for Freedom,"

Government of Canada. September, 15, 2015. https://www.canada.ca/en/news/archive/2015/09/battle-britain-fighting-freedom.html.

Gray, J., *Men Are from Mars, Women Are from Venus: Practical Guide for Improving Communication* (New York, Harper Collins, 1992).

Griffiths, C. and Buttery, T. "The World's Oldest Center of Learning," BBC. February, 24, 2022. https://www.bbc.com/travel/article/20180318-the-worlds-oldest-centre-of-learning.

Griswold, E., "How 'Silent Spring' Ignited the Environmental Movement," *The New York Times*. September, 21, 2012. https://www.nytimes.com/2012/09/23/magazine/how-silent-spring-ignited-the-environmental-movement.html.

Grogan, K., "How the Entire Scientific Community Can Confront Gender Bias in the Workplace." *Nature Ecology & Evolution*, 3 (2019: 3–6).

Hafner, K. and Scharf, A., "Female Scientists Who Worked on A-bomb Mostly Absent from 'Oppenheimer,'" *The Washington Post*. August, 10, 2023. https://www.washingtonpost.com/science/2023/08/10/oppenheimer-manhattan-project-women-scientists-bomb/.

Hall, R. and Sandler, B., *The Classroom Climate: A Chilly One for Women?* (Washington DC, Project on the Status and Education of Women, Association of American Colleges, 1982).

Hamblin, J., "No Doctor Should Work 30 Straight Hours Without Sleep," *The Atlantic*. December, 16, 2016. https://www.theatlantic.com/health/archive/2016/12/no-doctor-should-work-30-straight-hours/510395/.

Hammond, A., Matulevich, E., Beegle, K. et al., *The Equality Equation: Advancing the Participation of Women and Girls in STEM* (Washington DC, World Bank, 2020).

Handwerk, B., "Like Tiny Scientists, Babies Learn Best by Focusing on Surprising Objects," *Smithsonian*. April, 2, 2015. https://www.smithsonianmag.com/science-nature/tiny-scientists-babies-learn-best-focusing-surprise-objects-180954839/.

Hango, D., "Harassment and Discrimination Among Faculty and Researchers in Canada's Postsecondary Institutions," Statistics Canada. July, 16, 2021. https://www150.statcan.gc.ca/n1/pub/75-006-x/2021001/article/00006-eng.htm.

Hansler, J., "Trump's Comments on Former Ukraine Ambassador Raise Further Questions," CNN. September, 25, 2019. https://www.cnn.com/2019/09/25/politics/us-ambassador-ukraine-trump-call/index.html.

REFERENCES

Hechtman, L., Moore, N., Schulkey, C. et al., "NIH Funding Longevity by Gender." *Proceedings of the National Academy of Sciences*, 115, 31 (2018: 7943–7948).

Helba, C., Porter, A., Nicholson, S. et al., "Women Among Physics and Astronomy Faculty. Results from the 2018 Academic Workforce Survey," Education Resources Information Center. December, 2019. https://www.aip.org/statistics/data/women/faculty.

Henley, J., "Female-led Countries Handled Coronavirus Better, Study Suggests," *The Guardian*. August, 18, 2020. https://www.theguardian.com/world/2020/aug/18/female-led-countries-handled-coronavirus-better-study-jacinda-ardern-angela-merkel.

Henville, L., "How to Write a Compelling Letter of Reference," *University Affairs*. June, 17, 2021. https://www.universityaffairs.ca/career-advice/ask-dr-editor/how-to-write-a-compelling-letter-of-reference/.

Hewitt, K. and Smolina, A., "CanPhysCounts: Canada's First National EDI Survey of the Physics Community," CanPhysCounts. June, 9, 2021. https://www.canphyscounts.ca/_files/ugd/038a80_64940c29c9924d1cb25de0292d8c8935.pdf.

Higgins, C., "The Age of Patriarchy: How an Unfashionable Idea Became a Rallying Cry for Feminism Today," *The Guardian*. June, 22, 2018. https://www.theguardian.com/news/2018/jun/22/the-age-of-patriarchy-how-an-unfashionable-idea-became-a-rallying-cry-for-feminism-today.

Higgitt, R., "The trouble with 'science,'" *The Guardian*. February, 26, 2013. https://www.theguardian.com/science/the-h-word/2013/feb/26/trouble-science-scientists.

Hogenboom, M., "The Gender Biases that Shape our Brains," BBC. May, 27, 2021. https://www.bbc.com/future/article/20210524-the-gender-biases-that-shape-our-brains.08/04/2022.

Holloway, M., "Mary Leakey: Unearthing History." *Scientific American*, 271, 4 (1994: 37–40).

Holmes, R., "The Royal Society's Lost Women Scientists," *The Guardian*. November, 21, 2010. https://www.theguardian.com/science/2010/nov/21/royal-society-lost-women-scientists.

Holmes, R., *The Age of Wonder: How the Romantic Generation Discovered the Beauty and the Terror of Science* (New York, Pantheon Books, 2008).

Hoopes, L., "Nancy Hopkins' Keynote Speech Shockers," *Nature*. April, 1, 2011. https://www.nature.com/scitable/forums/women-in-science/nancy-hopkins-keynote-speech-shockers-19135206/.

213

Houser, K., "What Are the Key Challenges Facing Women in Academia?" Methodspace. March, 8, 2019. https://www.methodspace.com/blog/what-are-the-key-challenges-facing-women-in-academia.

Howard, J., "The History of the 'Ideal' Woman and Where that has Left Us," CNN. March, 9, 2018. https://www.cnn.com/2018/03/07/health/body-image-history-of-beauty-explainer-intl/index.html.

Hughes, K., "Gender Roles in the 19th Century," British Library. May, 15, 2014. https://www.bl.uk/romantics-and-victorians/articles/gender-roles-in-the-19th-century.

Humphries, C., "Measuring Up," MIT Technology Review. August, 16, 2017. https://www.technologyreview.com/2017/08/16/149744/measuring-up/#:~:text=She%20famously%20took%20a%20tape,3%2C000%20to%206%2C000%20square%20feet).

Hunt, K., "Lab Rats are Overwhelmingly Male, and That's a Problem," CNN. May, 14, 2021. https://www.cnn.com/2021/05/14/health/sex-biological-variable-research-science-drugs-scn/index.html.

Hutchison, P., "No" Now Really Does Mean "No", Queen's University. February, 2, 2010. https://www.queensu.ca/alumnireview/articles/2010-02-04/no-now-really-does-mean-no.

Ilham, T., "Fatima al-Fihri: Founder of the World's Oldest University," DW. May, 8, 2020. https://www.dw.com/en/fatima-al-fihri-founder-of-the-worlds-oldest-university/a-53371150.

Illes, J., "Women in Health, Science and Innovation are Collaborating Globally," The Conversation. March, 3, 2019. https://theconversation.com/women-in-health-science-and-innovation-are-collaborating-globally-111959.

Instead, J., Carstairs, C., and Hughes, K., "Before #Me Too: The Fight Against Sexual Harassment at Ontario Universities: 1979–1994." *Historical Studies in Education*, 33, 1 (2021: 1–21).

IPSOS, "Harcèlement Sexuel et Sexisme au Sein Du Monde Scientifique," IPSOS. March, 2023. https://www.ipsos.com/sites/default/files/ct/news/documents/2023-03/Ipsos-Fondation-Loreal-Harcelement-Sexisme-monde-scientifique.pdf.

Ireland, D., "Only About 1 in 5 Engineering Degrees Go to Women," The Conversation. June, 23, 2022. https://theconversation.com/only-about-1-in-5-engineering-degrees-go-to-women-185256.

Jackson, G., "The Female Problem: How Male Bias in Medical Trials Ruined Women's

Health," *The Guardian*. November, 13, 2019. https://www.theguardian.com/lifeandstyle/2019/nov/13/the-female-problem-male-bias-in-medical-trials.

Jacobi, T. and Schweers, D., "Justice, Interrupted: The Effect of Gender, Ideology and Seniority at Supreme Court Oral Arguments." *Virginia Law Review*, 103 (2017: 1379–1496).

Jessup-Anger, J. Lopez, E., and Koss, M., "History of Sexual Violence in Higher Education." *New Directions for Student Services: Special Issue: Addressing Sexual Violence in Higher Education and Student Affairs*, 2018, 161 (Spring 2018: 9–19).

Johnson, S, Hekman, D., and Chan, E., "If There's Only One Woman in Your Candidate Pool, There's Statistically No Chance She'll Be Hired," *Harvard Business Review*. April, 26, 2016. https://hbr.org/2016/04/if-theres-only-one-woman-in-your-candidate-pool-theres-statistically-no-chance-shell-be-hired.

Johnson, B., "CAMPUS CONFIDETIAL. University courtship is like an arcade game in which the rules are confusing and hazard warnings loom at every turn," *Macleans*. November, 9, 1992. https://archive.macleans.ca/article/1992/11/9/campus-confidetial.

Jones, C., "We Need to Put Back the Women Who Were Written Out of Science History," *The Independent*. December, 9, 2018. https://www.independent.co.uk/news/long_reads/women-written-out-science-history-ada-lovelace-caroline-herschel-a8666641.html.

Jordan, M., "First Woman President Named in Ivy League," *The Washington Post*. December, 7, 1993. https://www.washingtonpost.com/archive/politics/1993/12/07/first-woman-president-named-in-ivy-league/81e4083d-5f47-4eec-9343-d2ff187a4949/.

Kamenetz, A., "A Closer Look at Sexual Assaults on Campus. The History of Campus Sexual Assault," NPR. November, 30, 2014. https://www.npr.org/sections/ed/2014/11/30/366348383/the-history-of-campus-sexual-assault.

Kaplan, S. and Guarino, B., "Half of Women in Science Experience Harassment, a Sweeping New Report Finds," *The Washington Post*. June, 12, 2018. https://www.washingtonpost.com/news/speaking-of-science/wp/2018/06/12/half-of-women-in-science-experience-harassment-a-sweeping-new-report-finds/.

Kent, D., "Don't Be Born a Woman in Florence, if You Want Your Own Way," *History Today*, 70, 11 (2020). https://www.historytoday.com/archive/head-head/was-there-womens-renaissance.

Kerby-Fulton, K., Bugyis, K. and Van Engen, J. Intellectuals and Leaders in the Middle Ages (Cambridge, Cambridge University Press, 2020).

Kerkhoven, A., Russo, P., and Land-Zandstra, A., "Gender Stereotypes in Science Education Resources: A Visual Content Analysis," *PLoS ONE*, 11, 11 (2016).

Khazan, O., "The More Gender Equality, the Fewer Women in STEM," *The Atlantic*. February, 18, 2018. https://www.theatlantic.com/science/archive/2018/02/the-more-gender-equality-the-fewer-women-in-stem/553592/.

Kidron, R., Kaganovskiy, L., and Baron-Cohen, S., "Empathizing-systemizing Cognitive Styles: Effects of Sex and Academic Degree." *PLoS ONE*, 13, 3 (2018: 1–17).

Kirkpatrick, C. and Kanin, E., "Male Sex Aggression on a University Campus." *American Sociological Review*, XXII (1957: 52–58).

Kish, S., "Study Finds Brains of Girls and Boys Are Similar, Producing Equal Math Ability," Carnegie Mellon University. November, 8, 2019. https://www.cmu.edu/ni/news/2019/november/cantlon-equal-math-ability.html.

Kitchener, C. and Wong, A., "The Moral Catastrophe at Michigan State," *The Atlantic*. September, 12, 2018. https://www.theatlantic.com/education/archive/2018/09/the-moral-catastrophe-at-michigan-state/569776/.

Kleinman, D. L. and Thomas, J. (Eds.), *Preventing Sexual Harassment and Reducing Harm by Addressing Abuses of Power in Higher Education Institutions* (2023, Washington, DC, National Academies of Sciences, Engineering, and Medicine).

Knezz, S., "We Need STEM Mentors Who Can Reduce Bias and Fight Stereotypes," *Scientific American*. January, 19, 2021. https://www.scientificamerican.com/article/we-need-stem-mentors-who-can-reduce-bias-and-fight-stereotypes/.

Koichopolos, J., Ott, M., Maciver, A. et al., "Gender-based Differences in Letters of Recommendation in Applications for General Surgery Residency Programs in Canada." *Canadian Journal of Surgery*, 65, 2 (2022: E236–E241).

Kolb, B. and Gibb, R., "Brain Plasticity and Behaviour in the Developing Brain." *Journal of the Canadian Academy of Child and Adolescent Psychiatry*, 20, 4 (2011: 265–276).

Koren, M., "Lawrence Krauss and the Legacy of Harassment in Science," *The Atlantic*, October, 24, 2018. https://www.theatlantic.com/science/archive/2018/10/lawrence-krauss-sexual-misconduct-me-too-arizona-state/573844/.

Kosner, A., "The Mind at Work: Alison Gopnik on Learning More Like Children. Dropbox. October, 7, 2019. https://blog.dropbox.com/topics/work-culture/the-mind-at-work--alison-gopnik-on-learning-more-like-children.

Koss, M. and Rutherford, A., "What We Knew About Date Rape Then, and What We Know Now," *The Atlantic*. September, 26, 2018. https://www.theatlantic.com/ideas/archive/2018/09/what-surveys-dating-back-decades-reveal-about-date-rape/571330/.

Kowal, M., "Wally Funk Is Defying Gravity and 60 Years of Exclusion from Space," *The New York Times*. July, 26, 2021. https://www.nytimes.com/2021/07/19/science/wally-funk-jeff-bezos.html.

Kuo, M., "Recommendation Letters Reflect Gender Bias," *Science*. October, 3, 2016. https://www.science.org/content/article/recommendation-letters-reflect-gender-bias.

Lacey, J., "The Philosophical Baby by Alison Gopnik," *The Guardian*. August, 8, 2009. https://www.theguardian.com/books/2009/aug/08/philosophical-baby-alison-gopnik-review.

Langin, K., "Women Scientists Don't Get Authorship They Should, New Study Suggests," *Science*. June, 22, 2022b. https://www.science.org/content/article/women-scientists-don-t-get-authorship-they-should-new-study-suggests.

Langin, K., "A Sense of Belonging Matters. That's Why Academic Culture Needs to Change," *Science*. January, 16, 2019. https://www.science.org/content/article/sense-belonging-matters-s-why-academic-culture-needs-change.

Lanskey, S., "Can We Stop with the Sexist T-Shirts for Little Girls?" *Cosmopolitan*. August, 7, 2013. https://www.cosmopolitan.com/style-beauty/fashion/news/a14891/childrens-place-shirt-controversy-sexism/.

Lauzen, M., "The Celluloid Ceiling in a Pandemic Year: Employment of Women on the Top U.S. Films of 2021," The Center for the Study of Women in Television and Film. January, 1, 2022. https://womenintvfilm.sdsu.edu/wp-content/uploads/2022/01/2021-Celluloid-Ceiling-Report.pdf.

Lauzen, M., "Despite Rise in Female Directors Like Chloe Zhao and Emerald Fennell, Women Still Lack Opportunities in Hollywood," *Variety*. March, 25, 2021. https://variety.com/2021/voices/columns/women-opportunities-in-hollywood-1234937370/.

Lavy, V. and Sand, E., "On the Origins of Gender Gaps in Human Capital: Short- and Long-term Consequences of Teachers' Biases." *Journal of Public Economic*, 167 (2018: 263–279).

Lawson, K., Crouter, A., and McHale, S., "Links Between Family Gender Socialization Experiences in Childhood and Gendered Occupational Attainment in Young Adulthood." *Journal of Vocational Behavior*, 90 (2015: 26–35).

Lee, J. and McCabe, J., "Who Speaks and Who Listens: Revisiting the Chilly Climate in College Classrooms." *Gender and Society*, 35, 1 (2021: 32–60).

Lee, J., "6 Women Scientists Who Were Snubbed Due to Sexism," *National Geographic*. May, 19, 2013. https://www.nationalgeographic.com/culture/

article/130519-women-scientists-overlooked-dna-history-science?loggedin=true&rnd=1701150171684.

Lemelson-MIT, "Alice H. Parker, Central Heating with Natural Gas," Lemelson-MIT. No date. https://lemelson.mit.edu/resources/alice-h-parker.

Leppert, R. and Desilver, D., "118th Congress Has a Record Number of Women," Pew Research Center. January, 3, 2023. https://www.pewresearch.org/fact-tank/2023/01/03/118th-congress-has-a-record-number-of-women/.

Lerchenmueller, M. and Sorenson, O., "The Gender Gap in Early Career Transitions in the Life Sciences." *Research Policy*, 47, 6 (2018: 1007–1017).

Leslie, S., Cimpian, M. Meyer, M. et al., "Expectations of Brilliance Underlie Gender Distributions Across Academic Disciplines." *Science*, 347, 6219 (2015: 262–265).

Levack, B., "Witchcraft and the Law," In Levack, B. (ed.), *The Oxford Handbook of Witchcraft in Early Modern Europe and Colonial America* (Oxford, Oxford University Press, 2013).

Library of Parliament, "Women in the Parliament of Canada: 100 Years of Representation," Library of Parliament. December, 1, 2021. https://hillnotes.ca/2021/12/01/women-in-the-parliament-of-canada-100-years-of-representation/.

Lieberman, M., *Social: Why Our Brains Are Wired to Contact* (New York, Crown, 2013).

Lin, F., Oh, S., Gordon, L. et al., "Gender-based Differences in Letters of Recommendation Written for Ophthalmology Residency Applicants." *BMC Medical Education*, 19, 476 (2019).

Lindeman, T., "'Hate Is Infectious': How the 1989 Mass Shooting of 14 Women Echoes Today," *The Guardian*. December, 4, 2019. https://www.theguardian.com/world/2019/dec/04/mass-shooting-1989-montreal-14-women-killed.

Lindley, L., Clemens, S., Knibbs, S. et al., "Omnibus Survey of Pupils and Their Parents or Carers: Wave 5," Digital Education Resource Archive. March, 18, 2019. https://dera.ioe.ac.uk/id/eprint/32997/1/survey_of_pupils_and_their_parents_or_carers-wave_5.pdf.

Lindquist, C., McKay, T., and Witwer, A., *Detailed Findings from RTI's Study of L'Oreal USA's for Women in Science Program* (Research Triangle Park, RTI International, 2019).

Liu, K. and Dipietro Mager, N., "Women's Involvement in Clinical Trials: Historical Perspective and Future Implications," *Pharmacy Practice*, 14,1 (2016: 708).

Lowe, D., "Graduate Abuse," *Science*. October, 28, 2019. https://www.science.org/content/blog-post/graduate-abuse.

Maron, D., "Sexual Harassment Remains Common in the Sciences," *Scientific American*. June, 12, 2018. https://www.scientificamerican.com/article/sexual-harassment-remains-common-in-the-sciences/.

Maas, M., "Toys Began Being Marketed for Gender in the 1940s, Now a Movement Is Pushing Back," CNN. December, 19, 2019. https://www.cnn.com/2019/12/19/health/toys-gender-conversation-wellness/index.html.

Maglaty, J., "When Did Girls Start Wearing Pink?" *Smithsonian*. April, 7, 2011. https://www.smithsonianmag.com/arts-culture/when-did-girls-start-wearing-pink-1370097/.

Malkiel, N., *Keep the Damned Women Out: The Struggle for Coeducation* (Princeton, Princeton University Press, 2016).

Markel, H., *The Secret of Life: Rosalind Franklin, James Watson, Francis Crick, and the Discovery of DNA's Double Helix* (New York, W.W. Norton Company, 2021).

Markusoff, J., "Danielle Smith's UCP Base Has Big Demands. She's Wary of Going Quite That Far," CBC. November, 5, 2023. https://www.cbc.ca/news/canada/calgary/danielle-smith-ucp-convention-parental-rights-analysis-1.70191.

Martinez, A. and Christnacht, C., "Women Making Gains in STEM Occupations but Still Underrepresented," United States Census Bureau. January, 26, 2021. https://www.census.gov/library/stories/2021/01/women-making-gains-in-stem-occupations-but-still-underrepresented.html.

Mason, M., Wolfinger, N., and Goulden, M., *Do Babies Matter? Gender and Family in the Ivory Tower* (New Brunswick, Rutgers University Press, 2013a).

Mason, M., "The Baby Penalty," The Chronicle of Higher Education. August, 5, 2013b. https://www.chronicle.com/article/the-baby-penalty/.

Mauvais-Jarvis, F., Bairey Merz, N., Barnes, P. et al., "Sex and Gender: Modifiers of Health, Disease, and Medicine." *Lancet*, 396, 10250, (2020: 565–582).

Mayo Clinic Staff, "Children and Gender Identity: Supporting Your Child," Mayo Clinic. February, 23, 2022. https://www.mayoclinic.org/healthy-lifestyle/childrens-health/in-depth/children-and-gender-identity/art-20266811.

McElvery, R., "3 Questions: Nancy Hopkins on Improving Gender Equality in Academia," MIT. September, 30, 2020. https://news.mit.edu/2020/3-questions-nancy-hopkins-improving-gender-equality-in-academia-0930.

McGill Reporter Staff, "Women Lead 20 Percent of World's Top Universities for First Time," McGill. March, 8, 2021. https://reporter.mcgill.ca/women-lead-20-per-cent-of-worlds-top-universities-for-first-time/.

McKenna, M., "Canada's First (and Female) Science Minister Is a Badass," *National Geographic*. November, 5, 2015. https://www.nationalgeographic.com/science/article/canadas-first-and-female-science-minister-is-a-badass.

McKenzie, J., *Pauline Jewett: A Passion for Canada* (Montreal, McGill-Queen's University Press, 2014).

McKie, R., "Rachel Carson and the Legacy of Silent Spring," *The Guardian*. May, 27, 2012. https://www.theguardian.com/science/2012/may/27/rachel-carson-silent-spring-anniversary.

McKie, R., "Quantum Man: Richard Feynman's Life in Science by Lawrence M Krauss Review," *The Guardian*. May, 15, 2011. https://www.theguardian.com/science/2011/may/15/quantum-man-richard-feynman-review.

McLean, D., "Hateful Surgeons' Hall Riots that Opposed Women Doctors recalled 150 years on," *The Scotsman*. November, 17, 2020. https://www.scotsman.com/heritage-and-retro/heritage/hateful-surgeons-hall-riots-that-opposed-women-doctors-recalled-150-years-on-3038791.

McLeod, S., "Female Medical Students Sidelined but Not Out," *The Kingston Whig Standard*. July, 5, 2021. https://www.thewhig.com/opinion/columnists/female-medical-students-sidelined-but-not-out.

Medin, D., Lee, C., and Bang, M., "Point of View Affects How Science Is Done," *Scientific American*, October, 1, 2014. https://www.scientificamerican.com/article/point-of-view-affects-how-science-is-done/#:~:text=They%20influence%20what%20we%20choose,Evolutionary%20biology%20is%20one%20example.

Menasce Horowitz, J., "Most Americans See Value in Steering Children Toward Toys, Activities Associated with Opposite Gender," Pew Research Center. December, 19, 2017. https://www.pewresearch.org/fact-tank/2017/12/19/most-americans-see-value-in-steering-children-toward-toys-activities-associated-with-opposite-gender/.

Merritt, S., "Unwell Women by Elinor Cleghorn Review – Battle for the Female Body," *The Guardian*. June, 7, 2021. https://www.theguardian.com/books/2021/jun/07/unwell-women-by-elinor-cleghorn-review-battle-for-the-female-body.

Mervis, J., "Can Indigenous Knowledge and Western Science Work Together? New Center Bets Yes," *Science*. October, 25, 2023a. https://www.science.org/content/article/can-indigenous-knowledge-and-western-science-work-together-new-center-bets-yes.

Mervis, J., "Women Leaders at Six Top Research Universities Urge More Diversity in Semiconductor Workforce," *Science*. October, 2023b. https://www.science.

org/content/article/women-leaders-six-top-research-universities-urge-more-diversity-semiconductor-workforce#:~:text=%E2%80%9CIf%20we%20make%20things%20better,with%20Sian%20Beilock%2C%20president%20of.

Meyer, M., Cimpian, A., and Leslie, S., "Women Are Underrepresented in Fields Where Success Is Believed to Require Brilliance." *Frontiers in Psychology*, 6 (2015: 235).

Microsoft, *"Closing the STEM Gap: Why STEM Classes and Careers Still Lack Girls and What We Can Do About It,"* Microsoft. March, 13, 2018. https://query.prod.cms.rt.microsoft.com/cms/api/am/binary/RE1UMWz.

Miller, D., Nolla, K., Eagly, A. et al., "The Development of Children's Gender-Science Stereotypes: A Meta-analysis of 5 Decades of U.S. Draw-A-Scientist Studies." *Child Development*, 89, 6 (2018: 1943–1955).

Miller, D., Eagly, A., and Linn, M., "Women's Representation in Science Predicts National Gender-Science Stereotypes: Evidence From 66 Nations." *Journal of Educational Psychology*, 107, 3 (2015: 631–644).

Mitchell, A., "The Curious, Extraordinary Life of Anne Innis Dagg," *Canadian Geographic*. August, 13, 2019. https://canadiangeographic.ca/articles/the-curious-extraordinary-life-of-anne-innis-dagg/.

Moore, A., "Character Traits: Scientific Virtue." *Nature*, 532, 139 (2016).

Moore, L., *Women Before the Court: Law and Patriarchy in the Anglo-American World, 1600–1800* (Manchester, Manchester University Press, 2019).

Morgan, A., Way, S., Hoefer, M. et al., "The Unequal Impact of Parenthood in Academia," Science Advances. February, 24, 2021. https://www.science.org/doi/10.1126/sciadv.abd1996.

Morgan, W., Elder, K., and King, E., "The Emergence and Reduction of Bias in Letters of Recommendation." *Journal of Applied Social Psychology*, 43, 11 (2013: 2297–2306).

Mortillaro, N., "Where Are All the Black Astronomers and Physicists? Racism, Isolation Keeping Many Away," CBC. June, 22, 2022. https://www.cbc.ca/news/science/black-astronomers-canada-1.6494702.

Mortillaro, N., "How Historical Racism in Science Continues to Shape the Black Experience," CBC. August, 13, 2021. https://www.cbc.ca/radio/quirks/quirks-quarks-black-in-science-1.5909184.

Moss-Racusin, C., Dovidio, J., Brescoll, V. et al., "Science Faculty's Subtle Gender Biases Favor Male Students." *Journal of the Proceedings of the National Academy of Sciences USA*, 109, 41 (2012: 16474–16479).

Munroe, I., "Academia Has a Harassment Problem, StatsCan Study Finds," University Affairs. August, 13, 2021. https://www.universityaffairs.ca/news/news-article/academia-has-a-harassment-problem-statscan-study-finds/.

Myers, K., Tham, W., Yin, Y. et al., "Unequal Effects of the COVID-19 Pandemic on Scientists." *Nature Human Behaviour,* 8 (2020: 880–883).

Napp, C. and Breda, T., "The Stereotype that Girls Lack Talent: A Worldwide Investigation." Science Advances. March, 9, 2022. https://www.science.org/doi/10.1126/sciadv.abm3689.

Natarajan, P., "The Myth of the 'Jewish Mother of the Bomb,'" CNN. August, 9, 2013. https://www.cnn.com/2013/08/09/opinion/natarajan-mother-bomb.

National Academies of Sciences, Engineering, and Medicine, *Sexual Harassment of Women: Climate, Culture, and Consequences in Academic Sciences, Engineering, and Medicine* (Washington DC, The National Academies Press, 2018).

National Center for Education Statistics, "Status and Trends in the Education of Racial and Ethnic Groups," NCES. February, 2019. https://nces.ed.gov/programs/raceindicators/indicator_reg.asp.

National Center for Science and Engineering Statistics, *Women, Minorities, and Persons with Disabilities in Science and Engineering: 2021* (Virginia, National Science Foundation, 2021).

National Science Foundation, *Sexual Assault/Harassment Prevention and Response (SAHPR), Final Report* (Virginia, National Science Foundation, 2022).

Neklason, A., "Elite-College Admissions Were Built to Protect Privilege," *The Atlantic.* March, 18, 2019. https://www.theatlantic.com/education/archive/2019/03/history-privilege-elite-college-admissions/585088/.

Neubauer, S., Hublin, J., and Gunz, P., "The Evolution of Modern Human Brain Shape." Science Advances. January, 24, 2018. https://www.science.org/doi/10.1126/sciadv.aao5961.

Neukam, S., "Lawmakers Target 'Tech Bro Culture' in Financial Sector with Diversity Push," *The Hill.* June, 30, 2022. https://thehill.com/homenews/house/3543277-lawmakers-target-tech-bro-culture-in-financial-sector-with-diversity-push/.

Nobles, M., Womack, C., Wonkam, A. et al., "Racism: Overcoming Science's Toxic Legacy," *A Nature Special Issue* October, 20, 2022. https://www.nature.com/immersive/d42859-022-00031-8/index.html.

Nosek, B., Smyth, F., Sriram, N. et al., "National Differences in Gender–Science Stereotypes Predict National Sex Differences in Science and Math Achievement," *Proceedings of the National Academy of Sciences*, 106, 26 (2009: 10593–10597).

NPR, "How the Word 'Scientist' Came to Be," NPR. May, 21, 2010. https://www.npr.org/2010/05/21/127037417/how-the-word-scientist-came-to-be.

NPR, "Women's Space Dreams Cut Short, Remembered," NPR. May, 11, 2007. https://www.npr.org/2007/05/11/10125596/womens-space-dreams-cut-short-remembered.

O'Dea, R., Lagisz, M., Jennions, M. et al., "Gender Differences in Individual Variation in Academic Grades Fail to Fit Expected Patterns for STEM," Nature Communications. September, 25, 2018. https://www.nature.com/articles/s41467-018-06292-0.

Oliveira, O., Yifang, M., Woodruff, T. et al., "Comparison of National Institutes of Health Grant Amounts to First-Time Male and Female Principal Investigators," JAMA, 321, 9 (2019: 898–900).

Olsen, A., "College Classrooms Are Still Chilly for Women, as Men Speak More," SociologyDartmouth. January, 28, 2021. https://sociology.dartmouth.edu/news/2021/01/college-classrooms-are-still-chilly-women-men-speak-more.

Orben, A., Tomova, L., and Blakemore, S., "The Effects of Social Deprivation on Adolescent Development and Mental Health." Lancet Child & Adolescent Health, 4 (2020: 634–640).

Oreskes, N., "Racism and Sexism in Science Haven't Disappeared," Scientific American. October, 1, 2020. https://www.scientificamerican.com/article/racism-and-sexism-in-science-havent-disappeared/.

Ortenberg, R., "It's Nothing New: Sexism in the Lab," Science History Institute. June, 22, 2018. https://www.sciencehistory.org/distillations/its-nothing-new-sexism-in-the-lab.

Page, S., "She's Made 1,750 Wikipedia Bios for Female Scientists who Haven't Gotten Their Due," The Washington Post. October, 17, 2022. https://www.washingtonpost.com/lifestyle/2022/10/17/jess-wade-scientist-wikiepdia-women/.

Pak, E., "When Women Became Nuns to Get a Good Education," History. November, 1, 2018. https://www.history.com/news/women-education-medieval-nuns-church.

Paoletti, J., Pink and Blue: Telling the Boys from the Girls (Bloomington, Indiana University Press, 2012).

Park, R., "Why So Many Young Doctors Work Such Awful Hours." The Atlantic. February, 21, 2017. https://www.theatlantic.com/business/archive/2017/02/doctors-long-hours-schedules/516639/.

Parker, K., "What's Behind the Growing Gap Between Men and Women in College Completion?" Pew Research Center. November, 9, 2021. https://www.pewresearch.org/fact-tank/2021/11/08/whats-behind-the-growing-gap-between-men-and-women-in-college-completion/.

Parker, K., "Women in Majority-male Workplaces Report Higher Rates of Gender Discrimination," Pew Research Center. March, 7, 2018. https://www.pewresearch.org/fact-tank/2018/03/07/women-in-majority-male-workplaces-report-higher-rates-of-gender-discrimination/.

Parker, K. and Funk, C., "Gender Discrimination Comes in Many Forms for Today's Working Women," Pew Research Center. December, 14, 2017. https://www.pewresearch.org/short-reads/2017/12/14/gender-discrimination-comes-in-many-forms-for-todays-working-women/.

Parker, P., "The Historical Role of Women in Higher Education." *Administrative Issues Journal*, 5, 1 (2015: 3–14).

Parkes, E., "Scientific Progress Is Built on Failure," *Nature*. January, 10, 2019. https://www.nature.com/articles/d41586-019-00107-y.

PBS, "A Map of Gender Diverse Cultures," PBS. August, 12, 2015. https://www.pbs.org/independentlens/content/two-spirits_map-html/.

Pebdani, R., "Measuring the Toll of COVID-19 on Academic Parents and What We Can do About It. LSE. October, 31, 2022. https://blogs.lse.ac.uk/impactofsocialsciences/2022/10/31/measuring-the-toll-of-covid-19-on-academic-parents-and-what-we-can-do-about-it/.

Penner, M. and Smith-Carrier, T., "Gender Pay Gap: It's Roughly Half-a-million Dollars for Women Professors Across a Lifetime," University Affairs. July, 5, 2022. https://www.universityaffairs.ca/opinion/in-my-opinion/gender-pay-gap-its-roughly-half-a-million-dollars-for-women-professors-across-a-lifetime/.

Peters, D., "The Engineering Gender Gap: It's More Than a Numbers Game," University Affairs. January, 9, 2020. https://www.universityaffairs.ca/features/feature-article/the-engineering-gender-gap-its-more-than-a-numbers-game/.

Peters, D., "Pioneering Biologist Anne Innis Dagg Gets Her Due," University Affairs. June, 26, 2019. https://www.universityaffairs.ca/features/feature-article/pioneering-biologist-anne-innis-dagg-gets-her-due/.

Petroff, A., "The Exact Age When Girls Lose Interest in Science and Math," CNN Business. February, 28, 2017. https://money.cnn.com/2017/02/28/technology/girls-math-science-engineering/index.html.

Pettigrew, E., *The Silent Enemy: Canada and the Deadly Flu of 1918* (Saskatoon, Western Producer Prairie Books, 1983).

Pierson, R., *Canadian Women and the Second World War* (Ottawa, The Canadian Historical Association, 1983).

Pinkham, S., "How the Soviets Won the Space Race for Equality," *The New York Times*. July, 21, 2019.

Piper, M. and Samarasekera, I., *Nerve: Lessons on Leadership from Two Women Who Went First* (Toronto, ECW Press, 2021).

Pope, S., Koch, E., and Butts, S., "Strength in Numbers: Ranks of Female Professors Growing in Canada," Capital Currents. https://capitalcurrent.ca/strength-in-numbers-ranks-of-female-professors-growing-in canada/#:~:text=How%20 much%20have%20female%20positions%20risen%3F&text=Since%20 2014%2C%20the%20number%20of,the%20exception%20of%20full%20 professors.

Porter, A. and Ivie, R., "Women in Physics and Astronomy," AIP. https://www.aip.org/statistics/women.

Prasad, R., "How Trump Talks About Women - and Does it Matter?" BBC. November 28, 2019. https://www.bbc.com/news/world-us-canada-50563106.

Provincial Advisory Council on the Status of Women Newfoundland and Labrador, "*Sexual Violence in the Canadian Post-Secondary Education Context: Informing Policy, Enacting Change*," PACSW. November, 16, 2023. https://pacsw.ca/site/uploads/2023/11/Sexual-Violence-in-the-Canadian-Post-Secondary-Context-Website.pdf.

Rabesandratana, T., "U.K. Government Vow to End 'Woke' Science Draws Rebuke from Researchers." *Science*. October, 17, 2023. https://www.science.org/content/article/uk-government-vow-end-woke-science-draws-rebuke-researchers.

Rabin, R., "Health Researchers Will Get $10.1 Million to Counter Gender Bias in Studies," *The New York Times*. September, 23, 2014. https://www.nytimes.com/2014/09/23/health/23gender.html.

Randstad, "Women in STEM Careers-Where We Are in 2023?" Randstad. March, 3, 2023. https://www.randstad.ca/employers/workplace-insights/women-in-the-workplace/women-in-stem-where-we-are-now/.

Rawlins, S., "Turned Away at the Gates: The Struggle for Places for Women at Cambridge," *Varsity*. March, 15, 2019. https://www.varsity.co.uk/features/17317.

Reiches, M. and Richardson, S. "We Dug into Data to Disprove a Myth

About Women in STEM," *SLATE*. February, 11, 2020. https://slate.com/ technology/2020/02/women-stem-innate-disinterest-debunked.html?fbclid= IwAR1DHONpGwVXR9bEqf3LW2LmmfMdi33D4cXEwLhn1fA-N5n ESq6mIEAneB0.

Rippon, G., "A Gendered World Makes a Gendered Brain," Tedx Talk. June 11, 2020. https://www.google.com/search?q=gina+rippon+social+brain+ youtube&rlz=1C1GCEW_enCA925CA925&oq=gina+rippon+social+bra in+youtube&gs_lcrp=EgZjaHJvbWUyBggAEEUYOTIHCAEQIRigATIH CAIQIRigAdIBCjIwMjMyajBqMTWoAgCwAgA&sourceid=chrome&ie= UTF-8#fpstate=ive&vld=cid:76676cfd,vid:2s1hrHppl5E,st:475.

Rippon, G., *Gender and Our Brains. How New Neuroscience Explodes the Myths of the Male and Female Minds* (New York, Vintage Books, 2019).

Rippon, G., "The Trouble with Girls? Gina Rippon Asks Why Plastic Brains Aren't Breaking Through Glass Ceilings," The British Psychological Society. November, 7, 2016. https://www.bps.org.uk/psychologist/trouble-girls.

Rippon, G. "Mind the gender gap: the social neuroscience of belonging." *Frontiers in Human Neuroscience* vol 17 (2023). https://www.ncbi.nlm.nih.gov/pmc/ articles/PMC10116861/.

Ritzer, G. and Guppy, N., *Introduction to Sociology* (Los Angeles, Sage, 2014).

Rosenberg, Y., "How Anti-Semitism Shaped the Ivy League as We Know It," *The Atlantic*. September, 22, 2022. https://newsletters.theatlantic.com/ deep-shtetl/632c8ea068f61f0021dbfd41/mark-oppenheimer-interview-jewish-ivy-league-antisemitism/.

Ross, M., Glennon, B., and Murciano-Goroff, R., "Women Are Credited Less in Science Than Men." *Nature*, 608 (2022: 135–145).

Rossiter, M., *Women Scientists in America. Volume One: Struggles and Strategies to 1940* (Baltimore, The Johns Hopkins University Press, 1982).

Rubiano-Matulevich, E., Hammond, A., Beegle, K. et al., "Improving the Pathway From School to STEM Careers for Girls and Women," World Bank Blogs. February, 11, 2019. https://blogs.worldbank.org/opendata/ improving-pathway-school-stem-careers-girls-and-women.

Ruggeri, A., Walker, C., Lombrozo, T. et al., "How to Help Young Children Ask Better Questions?" *Frontiers in Psychology*, 11 (2021).

Russell, A. and Metcalf, H., "Transforming STEM Leadership Culture," Association for Women in Science. AWIS. 2019. https://awis.org/wp-content/ uploads/2019-Leadership-Report_FINAL_WEB.pdf.

Russell, C., "Confronting Sexual Harassment in Science," *Scientific American,* October 27, 2017. https://www.scientificamerican.com/article/confronting-sexual-harassment-in-science/.

Russo, N., "The Still-misunderstood Shape of the Clitoris," *The Atlantic.* March, 9, 2017. https://www.theatlantic.com/health/archive/2017/03/3d-clitoris/518991/.

Ryan, M., "To Advance Equality for Women, Use the Evidence," *Nature.* April, 19, 2022. https://www.nature.com/articles/d41586-022-01045-y.

Sadker, D., Sadker, M., and Zittleman, K., *Still Failing at Fairness: How Gender Bias Cheats Girls and Boys in School and What We Can Do About It* (New York, Scribner, 2009).

Sadker, M. and Sadker, D., *Failing at Fairness: How America's Schools Cheat Girls* (New York, Scribner, 1995).

Saewyc, E., "A Global Perspective on Gender Roles and Identity." *Journal of Adolescent Health,* 61, 4, (2017: S1–S2).

Saini, A., *Inferior: How Science Got Women Wrong-and the New Research That's Rewriting the Story* (Boston, Beacon Press, 2017).

Sample, I., "British Astrophysicist Overlooked by Nobels Wins $3m Award for Pulsar Work," *The Guardian.* September, 6, 2018. https://www.theguardian.com/science/2018/sep/06/jocelyn-bell-burnell-british-astrophysicist-overlooked-by-nobels-3m-award-pulsars.

Sample, I., "What Is This Thing We Call Science? Here's one definition..." *The Guardian.* March, 4, 2009. https://www.theguardian.com/science/blog/2009/mar/03/science-definition-council-francis-bacon.

Sampson, A., "Canada's Cold War Fallout Shelters Would Have Excluded Most of Us. These Women Had Other Plans," CBC. February, 28, 2023. https://www.cbc.ca/news/canada/nova-scotia/fallout-shelters-debunk-debert-protest-1.6759960.

Samson, N. and Shen, A., "A History of Canada's Full-time Faculty in Six Charts. A Look at Some UCASS Data From 1970 to 2016," University Affairs. March, 20, 2018. https://www.universityaffairs.ca/features/feature-article/history-canadas-full-time-faculty-six-charts/.

Saujani, R., "Closing the Gender Gap in Science and Technology," United Nations. February, 10, 2020. https://www.un.org/en/un-chronicle/closing-gender-gap-science-and-technology-0.

Schiebinger, L. (ed)., *Women and Gender in Science and Technology* (London, Routledge, 2014).

Schiebinger, L., "Women's Health and Clinical Trials." *The Journal of Clinical Investigation*, 112, 7 (2003: 973–977).

Schinske J., Perkins, H., Snyder, A., et al., "Scientist Spotlight Homework Assignments Shift Students' Stereotypes of Scientists and Enhance Science Identity in a Diverse Introductory Science Class," CBE-Life Sciences Education. October, 13, 2017. https://www.lifescied.org/doi/10.1187/cbe.16-01-0002.

Schmader, T., Whitehead, J., and Wysocki, V., "A Linguistic Comparison of Letters of Recommendation for Male and Female Chemistry and Biochemistry Job Applicants. *Sex Roles*, 57, 7–8, (2007: 509–514).

Schneegans, S., Lewis, and J., Straza, T. (eds). *UNESCO Science Report: The Race Against Time for Smarter Development* (Paris, UNESCO Publishing, 2021).

ScienceDaily, "Massive Study Reveals Few Differences Between Men's and Women's Brains," ScienceDaily. March, 29, 2021. https://www.sciencedaily.com/releases/2021/03/210325115316.htm.

Secord, J., "What Is the History of Science?" The British Academy. January, 14, 2021. https://www.thebritishacademy.ac.uk/blog/what-is-the-history-of-science/.

Secord, J., "Mary Somerville's Vision of science." *Physics Today*, 71, 1, (2018: 46–52).

Sheltzer, J. and Smith, J., "Elite Male Faculty in the Life Sciences Employ Fewer Women." *Proceedings of the National Academy of Sciences* 111, 28 (2014: 10107–10112).

Silicon Valley Bank Financial Group, "Women in Technology Leadership 2019," Silicon Valley Bank. 2019. https://www.svb.com/globalassets/library/uploadedfiles/content/trends_and_insights/reports/women_in_technology_leadership/svb-suo-women-in-tech-report-2019.pdf.

Simion, F. and Di Giorgio, E., "Face Perception and Processing in Early Infancy: Inborn Predispositions and Developmental Changes," *Frontiers in Psychology*. July, 9, 2015. https://www.frontiersin.org/articles/10.3389/fpsyg.2015.00969/full.

Singh, I., "By the Numbers: Women in STEM: What Do the Statistics Reveal About Ongoing Gender Disparities?" *Yale Scientific*. November, 27, 2020. https://www.yalescientific.org/2020/11/by-the-numbers-women-in-stem-what-do-the-statistics-reveal-about-ongoing-gender-disparities/.

Smith, M., "U15 Leadership Remains Largely White and Male Despite 33 Years of Equity Initiatives," Academic Women's Association University of Alberta. June, 20, 2019. https://uofaawa.wordpress.com/2019/06/20/u15-leadership-remains-largely-white-and-male-despite-33-years-of-equity-initiatives/.

REFERENCES

Somerville, L. and Gruber, J., "Three Trouble Spots Facing Women in Science—and How We Can Tackle Them," *Science.* October, 16, 2020. https://www.science.org/content/article/three-trouble-spots-facing-women-science-and-how-we-can-tackle-them.

Sottile, Z., "This Ivy League School Will Welcome Its First Female President After More Than 250 Years," CNN. July, 30, 2022. https://www.cnn.com/2022/07/30/us/sian-beilock-dartmouth-new-president-trnd/index.html.

Spelke, E., "Sex Differences in Intrinsic Aptitude for Mathematics and Science? A critical review." *American Psychologist,* 60, 9 (2005: 950–958).

Spence, N., "Once There Were No Women at Varsity." *University of Toronto Monthly,* January, 1933.

Stack, M., "Why I'm Not Surprised Nobel Laureate Donna Strickland Isn't a Full Professor," The Conversation. October, 4, 2018. https://theconversation.com/why-im-not-surprised-nobel-laureate-donna-strickland-isnt-a-full-professor-104459.

Staff, "Celebrating 120 Years of Women at U of T. Lectures and Events Mark Persistent Struggle for Equality," University of Toronto Magazine. June, 8, 2004. https://magazine.utoronto.ca/campus/history/celebrating-120-years-of-women-at-u-of-t/.

Stahl, A. and Feigenson, L., "Observing the Unexpected Enhances Infants' Learning and Exploration." *Science,* 348, 6230 (2015: 91–94).

Staniscuaski, F., Reichert, F., Werneck, F. et al., "Impact of COVID-19 on Academic Mothers." *Science,* 368, 6492 (2020: 724).

Statistics Canada, "Number and Salaries of Full-time Teaching Staff at Canadian Universities," Statistics Canada. November, 1, 2023. https://www150.statcan.gc.ca/t1/tbl1/en/tv.action?pid=3710010801.

Statistics Canada, "Study: Harassment and Discrimination Among Faculty and Researchers in Canada's Postsecondary Institutions," Statistics Canada. July, 16, 2021. https://www150.statcan.gc.ca/n1/daily-quotidien/210716/dq210716c-eng.htm.

Statistics Canada, "How Far Have We Come? Representation of Women Among Full-time University Faculty," Statistics Canada. March 5, 2020. https://www150.statcan.gc.ca/n1/pub/11-627-m/11-627-m2020021-eng.htm.

Steffens, M., Jelenec, P., and Noack, P., "On the Leaky Math Pipeline: Comparing Implicit Math-gender Stereotypes and Math Withdrawal in Female and Male Children and Adolescents." *Journal of Educational Psychology,* 102, 4 (2010: 947–963).

Stephen, J., "Balancing Equality for the Post-War Woman: Demobilising Canada's Women Workers After World War Two." *Atlantis*, 32, 1 (2007: 125–135).

Sterling, A., Thompson, M., Wang, S. et al., "The Confidence Gap Predicts the Gender Pay Gap Among STEM Graduates," *Proceedings of the National Academies of Sciences of the United States of America*, 117, 48 (2020: 30303–30308).

Stoet G., Geary D. C. "The gender-equality paradox in science, technology, engineering, and mathematics education." *Psychological Science*, 29 (2018: 581–593).

Stroh, P., "Pandemic Threatens to Wipe Out Decades of Progress for Working Mothers," CBC. August, 17, 2020. https://www.cbc.ca/news/business/women-employment-covid-economy-1.5685463.

Sylvester, R., "John Glenn and the Sexism of the Early Space Program," *Smithsonian*. December, 14, 2016. https://www.smithsonianmag.com/history/even-though-i-am-girl-john-glenns-fan-mail-and-sexism-early-space-program-180961443/.

Telegraph Staff and Agencies, "Mothers Asked Nearly 300 Questions a Day, Study Finds," The Telegraph. March, 28, 2013. https://www.telegraph.co.uk/news/uknews/9959026/Mothers-asked-nearly-300-questions-a-day-study-finds.html.

The Canadian Press, "Provinces and Territories Endorse National Action Plan to End Gender-based Violence," *Toronto Star*. November, 9, 2022. https://www.thestar.com/news/canada/2022/11/09/provinces-and-territories-endorse-national-action-plan-to-end-gender-based-violence.html.

The Canadian Press, "All Female Faculty Get Raises at University of Guelph," CBC. June, 19, 2018. https://www.cbc.ca/news/canada/kitchener-waterloo/university-guelph-female-faculty-1.4713063.

The Center for Legislative Archives, "Benjamin Franklin's Anti-Slavery Petitions to Congress," National Archives. August, 12, 2019. https://www.archives.gov/legislative/features/franklin.

The Editors, "Clinical Trials Have Far Too Little Racial and Ethnic Diversity," *Scientific American*. September, 1, 2018. https://www.scientificamerican.com/article/clinical-trials-have-far-too-little-racial-and-ethnic-diversity/.

The European Space Agency, "Mercury 13," ESA. June, 16, 2013. https://www.esa.int/About_Us/ESA_history/50_years_of_humans_in_space/Mercury_13.

The Francis Crick Institute, "About Francis Crick," The Francis Crick Institute. No date. https://www.crick.ac.uk/about-us/our-history/about-dr-francis-crick.

REFERENCES

The Lyda Hill Foundation and The Geena Davis Institute on Gender in Media, "Portray Her: Representation of Women STEM Characters in the Media," See Jane. September, 25, 2018. https://seejane.org/wp-content/uploads/portray-her-full-report.pdf.

The Royal Society of Canada, *Impacts of COVID-19 Pandemic on Women in Canada* (Ottawa, The Royal Society of Canada, 2022).

The Times Higher Education, "Women Lead 20 Percent of World's Top Universities for First Time," *Times Higher Education*. March, 5, 2021. https://www.timeshighereducation.com/academic/press-releases/women-lead-20-percent-worlds-top-universities-first-time.

The University of Texas at Austin, Office of the President, "Lorene Lane Rogers," The University of Texas at Austin, Office of the President. https://president.utexas.edu/past-presidents/lorene-lane-rogers.

Thomason, M., "Development of Brain Networks in Utero: Relevance for Common Neural Disorders." *Biological Psychiatry*, 88, 1 (2020: 40–50).

Thornton, A., "Gender Equality in STEM Is Possible: These countries Prove It," World Economic Forum. March, 5, 2019. https://www.weforum.org/agenda/2019/03/gender-equality-in-stem-is-possible/.

Tomova, L., Andrews, J., and Blakemore, S., "The Importance of Belonging and the Avoidance of Social Risk Taking in Adolescence," *Developmental Review*, 61, 100981 (2021).

Trevisan, M., "The Long Walk to Progress," The Queen's University Journal. July, 26, 2005. https://www.queensjournal.ca/story/2005-07-26/long-walk-progress/.

Trix, F. and Psenka, "Exploring the Color of Glass: Letters of Recommendation for Female and Male Medical Faculty." *Discourse and Society*, 14: (2003: 191–220).

Trotman, A., "Why Don't European Girls Like Science or Technology?" Microsoft News. March, 1, 2017. https://news.microsoft.com/europe/features/dont-european-girls-like-science-technology/.

Truth and Reconciliation Commission of Canada, *Canada's Residential Schools: Missing Children and Unmarked Burials. Final Report of the Truth and Reconciliation Commission of Canada* (Montreal, McGill-Queen's University Press, 2015).

Tucker, I., "The Five: Unsung Female Scientists," *The Guardian*. June, 16, 2019. https://www.theguardian.com/science/2019/jun/16/the-five-unsung-female-scientists-overlook-credit-stolen-jean-purdy.

TVO Today, "Transcript: Stephen Lewis: The Great Debate," TVO. November, 21, 2013. https://www.tvo.org/transcript/2162445/stephen-lewis-the-great-debate.

UN News, "Girls' Performance in Maths 'Starting to Add Up to Boys', says UNESCO," United Nations. April, 27, 2022. https://news.un.org/en/story/2022/04/1117082.

UN News, "After Year of 'Trials, Tragedies and Tears', UN Chief Sends Message of Hope for 2021," United Nations. December, 28, 2020a. https://news.un.org/en/story/2020/12/1080962.

UN News, "Make This the Century of Women's Equality: UN Chief," United Nations. February, 27, 2020b. https://news.un.org/en/story/2020/02/1058271. 17/09/2022.

UN Women, "Facts and Figures: Women's Leadership and Political Participation," UN Women. September, 18, 2023. https://www.unwomen.org/en/what-we-do/leadership-and-political-participation/facts-and-figures.

UN Women, "Invest in Women and Girls as Science Entrepreneurs, Urge Global Leaders from the Generation Equality Action Coalition on Technology and Innovation," UN Women. February, 11, 2022. https://www.unwomen.org/en/news-stories/news/2022/02/invest-in-women-and-girls-as-science-entrepreneurs-urge-global-leaders-from-the-generation-equality-action-coalition-on-technology-and-innovation.

UNDP, *Tackling Social Norms. A Game Changer for Gender Inequalities*, (New York, United Nations Development Programme, 2020).

UNESCO, "Girl Trouble: Breaking Through the Bias in AI," UNESCO. March, 8, 2021a. https://en.unesco.org/girltrouble.

UNESCO, "Women a Minority in Industry 4.0 Fields," UNESCO. February, 25, 2021b. https://www.unesco.org/en/articles/women-minority-industry-40-fields.

UNESCO, "One in Three Researchers is a Woman," UNESCO. February, 11, 2021c. https://www.unesco.org/en/articles/one-three-researchers-woman#:~:text=These%20regions%20are%20home%20to,Paraguay%20and%20Uruguay%20(49%25).

UNESCO, *Artificial Intelligence and Gender Equality. Key Findings of UNESCO's Global Dialogue* (Paris, UNESCO, 2020).

UNESCO, *Cracking the Code: Girls' and Women's Education in Science, Technology, Engineering and Mathematics (STEM)* (Paris, UNESCO, 2017).

Universities Canada, "Universities Canada: Inclusive Excellence Principles," Universities Canada. October, 2017. https://www.univcan.ca/wp-content/uploads/2017/10/equity-diversity-inclusion-principles-universities-canada-oct-2017.pdf.

University of Toronto Libraries, "Equal Opportunities. A Seat at the Table,"

University of Toronto Libraries. No date. https://exhibits.library.utoronto. ca/exhibits/show/changemakers/equal-opportunities.

Uppal, S. and Hango, D., "Differences in Tenure Status and Feelings of Fairness in Hiring and Promotions Among Male and Female Faculty in Canadian Universities," Statistics Canada. September, 1, 2022. https://www150. statcan.gc.ca/n1/pub/75-006-x/2022001/article/00007-eng.htm.

Vijayaraghavan, R., Duran, K., and Ramirez, K., "It's Time for Science and Academia to Address Sexual Misconduct," *Scientific American*. December, 12, 2017. https://blogs.scientificamerican.com/voices/its-time-for-science-and-academia-to-address-sexual-misconduct/#:~:text=Sexual%20and%20 racial%20misconduct%20has,free%20of%20harassment%20and%20abuse.

Villmoare, B. and Grabowski, M., "Did the Transition to Complex Societies in the Holocene Drive a Reduction in Brain Size? A Reassessment of the DeSilva et al. (2021) Hypothesis," *Frontiers in Ecology and Evolution*, 10 (2022). https:// www.frontiersin.org/articles/10.3389/fevo.2022.963568/full.

Vogel, G., "Medical Education Must Include the Field's Nazi Past, Expert Panel Urges," *Science*. November, 10, 2023. https://www.science.org/content/ article/medical-education-must-include-field-s-nazi-past-expert-panel-urges#:~:text=All%20health%20care%20students%20worldwide,medical% 20education%20in%20the%20future.

Wall., K., "Persistence and Representation of Women in STEM Programs," Statistics Canada. May 2, 2019. https://www150.statcan.gc.ca/n1/pub/75-006-x/2019001/article/00006-eng.htm.

Wamsley, L. "A Guide to Gender Identity Terms." NPR. June, 2, 2021. https://www.npr.org/2021/06/02/996319297/gender-identity-pronouns-expression-guide-lgbtq.

Wang, L., Stanovsky, G., Weihs, L., et al., "Gender Trends in Computer Science Authorship." *Communications of the ACM*, 64, 3 (2021: 78–84).

Waxman, O., "#DistractinglySexy Trends in Response to Nobel Scientist's Sexist Remarks," *Time*. June, 12, 2015. https://time.com/3918909/distractingly-sexy-tim-hunt/.

Weitekamp, M. and Cochrane, D., "Remembering Geraldyn 'Jerrie' Cobb, Pioneering Woman Aviator," National Air and Space Museum. April, 18, 2019. https://airandspace.si.edu/stories/editorial/remembering-geraldyn-jerrie-cobb-pioneering-woman-aviator.

Weitekamp, M., "NASA's Early Stand on Women Astronauts: 'No Present Plans to Include Women on Space Flights,' National Air and Space

Museum. March, 17, 2016. https://airandspace.si.edu/stories/editorial/nasas-early-stand-women-astronauts-%E2%80%9Cno-present-plans-include-women-space-flights%E2%80%9D#:~:text=%E2%80%9CWe%20have%20no%20present%20plans,the%20beginnings%20of%20the%20second.

Wen, D., Khan, S., Xu, A. et al., "Characteristics of Publicly Available Skin Cancer Image Datasets: A Systematic Review." *Lancet Digital Health,* 4, 1(2021: E64–E74).

Wetzel, C., "No Nobel Prizes in Science Went to Women This Year, Widening the Awards' Gender Gap," *Smithsonian.* October, 8, 2021. https://www.smithsonianmag.com/smart-news/the-nobel-gender-gap-widens-as-no-women-awarded-science-prizes-180978835/.

Wilkes, R. and Ramos, H., "It's Time to Rethink Letters of Recommendation," University Affairs. December, 7, 2018. https://www.universityaffairs.ca/opinion/in-my-opinion/its-time-to-rethink-letters-of-recommendation/.

Williams, J. and K. Massinger, How Women Are Harassed Out of Science," *The Atlantic.* July, 25, 2016. https://www.theatlantic.com/science/archive/2016/07/how-women-are-harassed-out-of-science/492521/.

Williams, J., "The 5 Biases Pushing Women Out of STEM," *Harvard Business Review.* March, 24, 2015. https://hbr.org/2015/03/the-5-biases-pushing-women-out-of-stem.

Williams, N. and Paperny, A., "Discovery of Children's Remains Reopens Wounds Among Indigenous Survivors of Colonial Canadian Schools," Reuters. June, 2, 2021. https://www.reuters.com/world/americas/mass-grave-reopens-wounds-among-indigenous-survivors-colonial-canadian-school-2021-06-02/.

Wills, M., "The Bluestockings," JSTOR Daily. April, 4, 2019. https://daily.jstor.org/the-bluestockings/.

Witze, A., "Three Extraordinary Women Run the Gauntlet of Science—a Documentary. *Nature,* 583, 7814 (2020: 25–26).

Witze, A., "US Science Agency will Require Universities to Report Sexual Harassment," *Nature.* February, 8, 2018. https://www.nature.com/articles/d41586-018-01744-5.

Woetzel, J., "How Women Could Build Economies the Size of the U.S. and China Combined," *Fortune.* October, 15, 2015a. https://fortune.com/2015/10/15/women-gender-gap-economic-growth-u-s-china/.

Woetzel, J., Madgavkar, A., Ellingrud, K. et al., "How Advancing Women's Equality Can Add $12 Trillion to Global Growth," McKinsey Global Institute. September, 1, 2015b. https://www.mckinsey.com/featured-insights/

employment-and-growth/how-advancing-womens-equality-can-add-12-trillion-to-global-growth.

Wolfers, J., "Fewer Women Run Big Companies Than Men Named John," *The New York Times*. March, 2, 2015. https://www.nytimes.com/2015/03/03/upshot/fewer-women-run-big-companies-than-men-named-john.html.

Wong, K., "The Trailblazing Black Woman Chemist Who Discovered a Treatment for Leprosy," *Smithsonian*. March, 23, 2022. https://www.smithsonianmag.com/history/the-trailblazing-black-woman-chemist-who-discovered-a-treatment-for-leprosy-180979772/.

Wood, J., "3 Things to Know About Women in STEM," World Economic Forum. February, 11, 2020 https://www.weforum.org/agenda/2020/02/stem-gender-inequality-researchers-bias/.

Woods, P. and Walker, A., "The Representation of Blackness in Astronomy." *Nature Astronomy*, 6 (2022: 622–624).

Woolston, C., "'It's Like We're Going Back 30 Years': How the Coronavirus is Gutting Diversity in Science," *Nature*. July, 31, 2020. https://www.nature.com/articles/d41586-020-02288-3.

World Economic Forum, *Global Gender Gap Report 2022* (Geneva, World Economic Forum, 2022).

World Economic Forum, *Global Gender Gap Report 2021* (Geneva, World Economic Forum, 2021).

World Economic Forum, *Jobs of Tomorrow: Mapping Opportunity in the New Economy* (Geneva, World Economic Forum, 2020a.)

World Economic Forum, *The Future of Jobs Report 2020* (Geneva, World Economic Forum, 2020b).

World Health Organization, *Care for Child Development: Participant Manual* (Geneva, World Health Organization, 2012).

World Health Organization, "Gender and Health," World Health Organization. No date. https://www.who.int/health-topics/gender#tab=tab_1.

Yang, M., "Let Her Finish: Interruptions of Female Justices Led to New Supreme Court Rules," *The Guardian*. October, 15, 2021. https://www.theguardian.com/law/2021/oct/15/us-supreme-court-female-justices-interruptions-sonia-sotomayor.

Yong, E., "What We Learn From 50 Years of Kids Drawing Scientists." *The Atlantic*. March, 18, 2018. https://www.theatlantic.com/science/archive/2018/03/what-we-learn-from-50-years-of-asking-children-to-draw-scientists/556025/.

Yong, E., "6-Year-Old Girls Already Have Gendered Beliefs About Intelligence,"

The Atlantic. January, 26, 2017. https://www.theatlantic.com/science/archive/2017/01/six-year-old-girls-already-have-gendered-beliefs-about-intelligence/514340/.

Yong, E., "XY Bias: How Male Biology Students See Their Female Peers," *The Atlantic.* February, 16, 2016. https://www.theatlantic.com/science/archive/2016/02/male-biology-students-underestimate-their-female-peers/462924/.

Zernike, K., *The Exceptions; Nancy Hopkins, MIT, and the Fight for Women in Science* (New York, Scribner, 2023).

Zhao, S., Setoh, P., Storage, D. et al., "The Acquisition of the Gender-Brilliance Stereotype: Age Trajectory, Relation to Parents' Stereotypes, and Intersections with Race/Ethnicity." *Child Development;* 93 (2022: e581–e597).

Zielinski, S., "Ten Historic Female Scientists You Should Know," *Smithsonian.* September, 19, 2011. https://www.smithsonianmag.com/science-nature/ten-historic-female-scientists-you-should-know-84028788/.

Zou, J. and Schiebinger, L., "AI Can Be Sexist and Racist—It's Time to Make it Fair." *Nature,* 559 (2018: 324–326).

Notes

1   Collier, 1974; Crosby, 1976; Beveridge, 1977; and Pettigrew, 1983
2   Duncan, 2003
3   Beveridge, 1977
4   Duncan, 2003
5   Collier, 1974; Pettigrew, 1983; and Duncan, 2003
6   Collier, 1974 and Duncan, 2003
7   Collier, 1974 and Pettigrew, 1983
8   Duncan, 2003
9   Pettigrew, 1983 and Duncan, 2003
10   Duncan, 2003
11   Crosby, 1976
12   Pettigrew, 1983
13   Collier, 1974; Pettigrew, 1983; and Duncan, 2003
14   Duncan, 2003
15   Duncan, 2003
16   Parkes, 2019
17   Duncan, 2003
18   Duncan, 2003
19   Duncan, 2003
20   Duncan, 2003
21   Duncan, 2003
22   Duncan, 2003
23   Duncan, 2003
24   Duncan, 2003
25   Duncan, 2003
26   Duncan, 2003

27 Duncan, 2003
28 Duncan, 2003
29 Duncan, 2003
30 Duncan, 2003
31 Duncan, 2003
32 Duncan, 2003
33 Duncan, 2003
34 AAUW, 2010 and Khazan, 2018
35 Rippon, 2019
36 Rippon, 2019
37 Rippon, 2016 and Cooke, 2022
38 Eliot, 2019
39 Thomason, 2020
40 Budday et al., 2015; Rippon, 2019; and CNN, 2014
41 Rippon, 2019
42 Eliot et al., 2021
43 Rippon, 2019
44 Kolb et al., 2011
45 Rippon, 2019
46 Angier, 2012
47 Spelke, 2005
48 CBC, 2019 and Rippon, 2019
49 Rippon, 2019
50 Rippon, 2019
51 Rippon, 2019 and Eliot, 2019
52 Rippon, 2019
53 Rippon, 2019
54 Eliot, 2021 and Science Daily, 2021
55 Cooke, 2022 and Gray, 1992
56 Eliot, 2018
57 Rippon, 2019
58 Kolb et al., 2011; Rippon, 2019; and Fox, 2019
59 Simion et al., 2015
60 Rippon, 2019
61 PBS, 2015
62 Asmelash, 2020
63 Hogenboom, 2021

64 World Health Organization, No date
65 World Health Organization, No date
66 Rippon, 2019
67 Fromson, 2011
68 Paoletti, 2012
69 Maglaty, 2011
70 Hogenboom, 2021
71 Rippon, 2019
72 Rippon, 2019
73 Eliot, 2009
74 Mayo Clinic Staff, 2022
75 Rippon, 2019
76 Mayo Clinic Staff, 2022
77 Hogenboom, 2021
78 Hogenboom, 2022
79 Eliot, 2009 and Rippon, 2019
80 Rippon, 2019
81 World Health Organization, 2012
82 Gopnik, 2012
83 Stahl et al., 2015
84 Handwerk, 2015
85 Lacey, 2009
86 Gopnik, 2011
87 Moore, 2016
88 Wamsley, 2021
89 Busch, 2020
90 Ruggeri et al., 2021
91 Telegraph Staff and Agencies, 2013
92 Elsworthy, 2017
93 Rippon, 2019
94 Rippon, 2019
95 Rippon, 2019
96 Bian et al., 2017
97 Rippon, 2019
98 Alam et al., 2020
99 O'Dea, R., Lagisz, M., Jennions, M. et al., "Gender Differences in Individual Variation in Academic Grades Fail to Fit Expected Patterns for STEM,"

Nature Communications. September, 25, 2018. https://www.nature.com/articles/s41467-018-06292-0.

100 UNESCO, 2017 and UN News, 2022
101 Bian et al., 2017
102 Kerkhoven et al., 2016
103 Nosek et al., 2009
104 Yong, 2017; Bian et al., 2018; and Rippon, 2019
105 Alam et al., 2020
106 Steffens et al., 2010
107 Rippon, 2019
108 Gopnik, 2011
109 Gopnik et al., 2001 and Kosner, 2019
110 UNDP, 2020
111 Bailey et al., 2022
112 Rippon, 2019
113 Goldenberg, 2005
114 Miller et al., 2015
115 Gjersoe, 2018
116 Miller et al., 2018
117 Yong, 2018
118 Yong, 2018
119 Charlesworth, 2019
120 Yong, 2017
121 Lindley et al., 2019
122 Alam, 2020
123 Bushwick et al., 2023
124 Accenture, 2017
125 Kish, 2019
126 Rippon, 2019
127 Hogenboom, 2022
128 Rippon, 2019
129 Zielinski, 2011
130 Rippon, 2019
131 Alam et al., 2020
132 Lavy et al., 2018
133 Lavy et al., 2018
134 Alam et al., 2020

135 Hammond et al., 2020
136 The Lyda Hill Foundation and The Geena Davis Institute on Gender in the Media, 2018
137 The Lyda Hill Foundation and The Geena Davis Institute on Gender in the Media, 2018
138 The Lyda Hill Foundation et al., 2018
139 Kent, 2020
140 Wills, 2019
141 Hughes, 2014
142 Saini, 2017
143 Cole, 2020
144 Ritzer et al., 2014
145 Neubauer et al., 2018
146 Villmoare et al., 2022
147 Higgins, 2018
148 Rippon, 2019
149 Howard, 2018
150 Moore, 2019
151 Bierschbach, 2019
152 Burgen, 2022 and Levack, 2013
153 Rippon, 2019
154 Holmes, 2010
155 Lee, 2013 and Jones, 2018
156 Schiebinger, 2014 and Dominus, 2019
157 NPR, 2010 and Rippon, 2019
158 Baldwin, 2014
159 Rippon, 2019 and Baldwin, 2014
160 Campbell, 1924
161 UNESCO, 2017
162 Rippon, 2019
163 Saewyc, 2017
164 Rippon, 2019
165 Maas, 2019
166 Hogenboom, 2021
167 Lawson et al., 2015
168 Menasce Horowitz, 2017
169 Maas, 2017

170 Lansky, 2013
171 Common Sense Media, 2017
172 Lauzen, 2021 and Lauzen, 2022
173 World Economic Forum, 2021
174 Wolfers, 2015
175 Lieberman, 2013
176 D'Mello, 2019
177 Rippon, 2019
178 D'Mello, 2019
179 Rippon, 2019
180 Lieberman, 2013 and Rippon, 2019
181 Orben et al., 2020 and Tomova et al., 2021
182 Orben et al., 2020
183 Rippon, 2019
184 Rippon, 2019
185 Rippon, 2019
186 Rippon, 2019
187 Rippon, 2019 and Lieberman, 2013
188 Eisenberger et al., 2003, 2005 and Rippon, 2019
189 Higgitt, 2013
190 Sample, 2009
191 Medin et al., 2014
192 British Science Council, 2016
193 Holmes, 2010
194 Holmes, 2008
195 Holmes, 2008
196 Holmes, 2010
197 Secord, 2018
198 Secord, 2021
199 Bernstein, 2015
200 Donald, 2015
201 Chung, 2019b
202 The Center for Legislative Archives, 2019
203 Canada's Aviation Hall of Fame, 2023
204 CBC, 2017
205 Canada's Aviation Hall of Fame, 2023 and Government of Canada, 2020
206 Government of Canada, 2015

207 CBC, 2017 and Government of Canada, 2020
208 Rippon, 2019
209 UN News, 2020b
210 Zhao et al., 2022
211 Leslie et al., 2015; Meyer et al., 2015; Bian et al., 2017, 2018b
212 Yong, 2016
213 Leslie et al., 2015
214 Napp et al., 2022
215 Aubourg, 2022
216 Trotman, 2017
217 Microsoft, 2018
218 George, 1915
219 Crim, 2005
220 Merritt, 2021
221 Cleghorn, 2021
222 Rippon, 2019
223 Baron-Cohen et al., 2005
224 Rippon, 2019
225 Barnett et al., 2017
226 Focquaert et al., 2007
227 Kidron et al., 2018
228 Rippon, 2019
229 Rippon, 2019
230 AAUW, No date
231 Rippon, 2019
232 Rippon, 2019
233 Rippon, 2019
234 Lemelson-MIT, No date
235 Trotman, 2017 and Petroff, 2017
236 Trotman, 2017
237 Bayer Corporation, 2012
238 Bayer, 2010
239 Bayer Corporation, 2012
240 Rubiano-Matulevich et al., 2019
241 Hammond et al., 2020
242 Rubiano-Matulevich et al., 2019
243 Reiches et al., 2020.

244 Stoet et al.
245 Reiches et al., 2020; and Rippon, 2023.
246 Peters, 2020
247 Wall, 2019
248 National Center for Education Statistics, 2019
249 National Center for Science and Engineering Statistics, 2021
250 Saujani, 2020
251 Ireland, 2022
252 Porter et al., No date
253 Wood, 2020
254 Kerby-Fulton et al., 2020
255 Griffiths et al., 2022
256 Ilham, 2020
257 Pak, 2018
258 Baston, 2008 and Darby, 2019
259 BBC, 2019
260 Rawlins, 2019
261 BBC, 2019
262 Anonymous, 2020a
263 Staff, 2004
264 Engineering, 2015
265 Malkiel, 2016
266 Anonymous, 2019
267 Staff, 2004
268 TVO Today, 2013
269 Staff, 2004
270 Rosenberg, 2022
271 Gladwell, 2005
272 Neklason, 2019
273 Rosenberg, 2022
274 Gladwell, 2005
275 Rosenberg, 2022
276 Gladwell, 2005
277 Wilkes et al., 2018; Henville, 2021; and Kuo, 2016
278 Trix et al., 2003; Schmader et al., 2007; Morgan et al., 2013; Lin et al., 2019; and Koichopolos et al., 2022
279 Moss-Racusin et al., 2012

NOTES

280 Dutt et al., 2016
281 Langin, 2019
282 Rippon, 2019
283 Cohen, 2016
284 Lowe, 2019
285 Park, 2017
286 Hamblin, 2016
287 Park, 2017
288 Hall et al., 1982
289 BBC, 2019b
290 Markel, 2021
291 BBC, 2015 and Waxman, 2015
292 Sadker et al., 2009
293 Lee et al., 2021
294 Rippon, 2019
295 Jacobi et al., 2017
296 Yang, 2021
297 Olsen, 2021
298 Brown, 1989
299 Hutchison, 2010
300 Instead et al., 2021
301 UN Women, 2022
302 Barnes, 1996
303 Holloway, 1996
304 McKie, 2012
305 Griswold, 2012
306 Des Jardins, 2011
307 Natarajan, 2013
308 Hafner et al., 2023
309 Page, 2022 and Ferguson, 2023
310 Brandt et al., 2020
311 Bowling et al., 1985
312 Knezz, 2021
313 Byars-Winston et al., 2019
314 Knezz, 2021 and Byars-Winston et al., 2019
315 Byars-Winston et al., 2019
316 Universities Canada, 2017

317 Saini, 2017
318 Illes, 2019
319 Illes, 2019
320 Sommerville et al., 2020
321 Parker, 2015
322 Ford, 1985
323 Parker, 2015
324 College of Medicine and Veterinary Medicine, 2018
325 McLean, 2020
326 College of Medicine and Veterinary Medicine, 2018
327 Ford, 1985
328 Williams et al., 2021
329 Truth and Reconciliation Commission of Canada, 2015
330 Ford, 1985
331 Trevisan, 2005
332 McLeod, 2021
333 Trevisan, 2005
334 Catudella, 1999
335 Ford, 1985
336 Ford, 1985
337 University of Toronto Libraries, No date
338 Samson et al., 2018
339 Parker, 2015
340 Rossiter, 1982; Dominus, 2019
341 Cohen, 2015
342 Colwell, 2020; Humphries, 2017; and Zernicke, 2023
343 Funk et al., 2018
344 Funk et al., 2018 and Russell et al., 2017
345 Parker et al., 2017
346 Illes, 2019
347 UNESCO, 2021
348 BBC, 2021 and France-Presse, 2021
349 Wang et al., 2021
350 Bello et al., 2021; Singh, 2020 and Russell et al., 2019
351 Statistics Canada, 2020
352 Anonymous, 2013
353 Bello et al., 2021
354 Oreskes, 2020 and Edge, 2020

355 Mortillaro, 2021
356 Mortillaro, 2021
357 Mortillaro, 2022
358 Woods et al., 2022
359 Collins et al., 2021; Myers et al., 2020; Staniscuaski et al., 2020; and Calaza et al., 2021
360 Jackson, 2019
361 Russo, 2017 and Cleghorn, 2021
362 Schiebinger, 2003
363 Jackson, 2019
364 Mauvais-Jarvis, 2020
365 Dubé et al., 2018
366 Liu et al., 2016
367 Hunt, 2021
368 The Editors, 2018 and Clayton et al., 2014
369 Jackson, 2019 and Rabin, 2014
370 Criado Perez, 2019b
371 Criado Perez, 2019a
372 UNESCO, 2021a
373 UNESCO, 2020
374 UNESCO, 2020
375 Zou et al., 2018
376 Wen et al., 2021
377 Buolamwini, 2019
378 Asare, 2018
379 Cohn, 2016
380 World Economic Forum, 2020a
381 European Institute for Gender Equality, 2017
382 Woetzel, 2015a and b
383 World Economic Forum, 2020a
384 Bello et al., 2021
385 Thornton, 2019
386 UNESCO, 2021c
387 Eveleth, 2013
388 Alam, 2020
389 Fry et al., 2021
390 Fry et al., 2021
391 Wall, 2019

392 Frank, 2019
393 Wall, 2019
394 Frank, 2019
395 Randstad, 2022
396 UNESCO, 2021c
397 Martinez et al., 2021
398 Ireland, 2022
399 Fouad et al., 2017
400 UNESCO, 2021c
401 Firth-Butterfield, 2021
402 Schneegans et al., 2021
403 Firth-Butterfield, 2021
404 Duke, 2018
405 UNESCO, 2021c
406 Deloitte AI Institute, 2020
407 UNESCO, 2021b
408 Silicon Valley Bank, 2019
409 UNESCO, 2021
410 Neukam, 2022
411 Colwell, 2020
412 Sheltzer et al., 2014 and Grogan, 2019
413 Sheltzer et al., 2014
414 Lindquist et al., 2019
415 Council of Canadian Academies, 2021
416 Becker et al., 2015
417 Callier, 2016
418 Johnson et al., 2016 and Moss-Racusin et al., 2012
419 Uppal et al., 2022
420 Cummings, 2020
421 Penner et al., 2022
422 Statistics Canada, 2023
423 Booker, 2020
424 Bueckert, 2016; CBC News, 2015; CBC News, 2017; and The Canadian Press, 2018
425 Caranci et al., 2017
426 Sterling et al., 2020
427 Binns, 2021

428 Caranci et al., 2017
429 Funk et al., 2018
430 Baker et al., 2019
431 Uppal et al., 2022
432 Caranci et al., 2017
433 Lerchenmueller et al., 2018
434 Uppal et al., 2022
435 Grogan, 2019
436 Colwell, 2020 and Oliveira et al., 2019
437 Grogan, 2019 and Hechtman, 2018
438 Grogan, 2019
439 Langin, 2022b; Ross et al., 2022
440 Grogan, 2019
441 Calisi et al., 2018
442 Cech et al., 2019
443 Frank, 2019
444 Mason et al., 2013a and 2013b
445 Mason, 2013b
446 Morgan et al., 2021
447 Fathima et al., 2020
448 Colwell, 2020
449 Colwell, 2020
450 Calisi, 2018
451 UN News, 2020; Stroh, 2020; and The Royal Society of Canada, 2022
452 Staniscuaski, 2020; CohenMiller et al., 2022; Davis et al., 2022; and Pebdani, 2022
453 Woolston, 2020
454 Grogan, 2019
455 Lee, 2013
456 Wong, 2022
457 Tucker, 2019
458 Burnell, 2023
459 Sample, 2018
460 UNESCO, 2021
461 Grogan, 2019; BBC, 2021
462 Wetzel, 2021
463 Gibney, 2018; Grogan, 2019; Wetzel, 2021; and Anonymous, 2018

464 Stack, 2018

465 McKenna, 2015

466 Anonymous, 2012

467 Anonymous, 2012 and Fitzpatrick, 2012

468 Peters, 2020

469 Government of Canada, 2023a

470 Pierson, 1983 and Stephen, 2007

471 Admin, 2020

472 Kirkpatrick et al., 1957

473 McElvery, 2020

474 Hoopes, 2011

475 The Francis Crick Institute, No date

476 NASEM, 2018

477 McKie, 2011 and Lincoln, 2018

478 Koren, 2018

479 Carstairs et al., 2021

480 Farley, 2017

481 Carstairs et al., 2021

482 Jessup-Anger, 2018

483 Koss et al., 2018

484 Kamenetz, 2014

485 Kamenetz, 2014

486 Carstairs et al., 2021

487 Peters, 2019; Deveau, 2022; and Mitchell, 2019

488 Carstairs et al., 2021

489 Johnson, 1992

490 Carstairs et al., 2021

491 Lindeman, 2019 and Chung, 2019

492 Kaplan et al., 2018

493 Williams et al., 2016; Vijayaraghavan et al., 2017; and IPSOS, 2023

494 The Canadian Press, 2022

495 Jessup-Anger, 2018

496 BBC, 2023

497 Asmelash, 2022

498 Russell, 2017

499 Cantor et al., 2020

500 NASEM, 2018

501 Statistics Canada, 2021
502 NASEM, 2018
503 Casarez et al., 2022
504 Kitchener et al., 2018
505 Casarez et. al, 2022
506 Cantor et al., 2020
507 Burczycka, 2020
508 NASEM, 2018
509 Munroe, 2021
510 NASEM, 2018; Corbyn, 2019
511 NASEM, 2018
512 Clancy et al., 2014
513 Clancy et al., 2017
514 National Science Foundation, 2022
515 Hango, 2021
516 Provincial Advisory Council on the Status of Women Newfoundland and Labrador, 2023
517 Cadloff, 2022
518 NASEM, 2018; Witze, 2018;
519 NASEM, 2018
520 NASEM, 2018
521 NASEM, 2018
522 Kleinman et al., 2023 and Batty, 2018
523 NASEM , 2018
524 Saini, 2017 and Rippon, 2019
525 Ford, 1985
526 McKenzie, 1999
527 The University of Texas at Austin, No date
528 Ford, 1985
529 Jordan, 1993
530 Coughlan, 2015
531 Sottile, 2022
532 Casey, 2022
533 The Times Higher Education, 2021
534 McGill Reporter Staff, 2021
535 Commodore, 2023
536 Gagliardi et al., 2017

537 Cafley, 2021
538 Cafley, 2021
539 Piper et al., 2021
540 Bray et al., 2020
541 Cafley, 2021
542 Smith, 2019
543 Bray et al., 2020
544 Government of Canada 2023 and Famous Five Foundation, no date
545 Government of Canada, 2023 and Famous5 Foundation, no date
546 Elections Canada, 2023
547 Bell, 2019
548 Bell, 2019
549 Sampson, 2023
550 UN Women, 2023
551 Deloitte, 2020
552 Henley, 2020
553 Cox, 2022
554 UN Women, 2023
555 Library of Parliament, 2021
556 Leppert et al., 2023
557 CBC, 2018
558 Funk, 2009
559 CBC Radio, 2018 and Weitekamp et al., 2019
560 NPR, 2007
561 Kowal, 2021
562 CBC Radio, 2018
563 Weitekamp, 2016; Sylvester, 2016; and The European Space Agency, 2013
564 CBC Radio, 2018; Weitekamp et al., 2019; Dejevsky, 2017; and Pinkham, 2019
565 Cox, 2022
566 Smith, 2019 and Cafley, 2021
567 Anonymous, 2020b; Mervis, 2023a and 2023b; and Vogel, 2023
568 AAU, 2023
569 Ball, 2023
570 Rabesandratana, 2023
571 Markusoff, 2023
572 Chapin, 2017
573 Penner et al., 2022